CAMBR CTION

Travel, Europe

This collection of narratives contains vivid accounts of the varied landscapes, built environment and customs encountered by eighteenth- and nineteenth-century travellers in the British Isles and Europe. Some were wealthy individuals on the Grand Tour, while others were travelling on business, for pleasure, in pursuit of better health, or simply to escape trouble at home.

Travels in Crete

Robert Pashley (1805–59) spent 1833–4 exploring Greece and Turkey as a Trinity College, Cambridge Travelling Fellow and contributor to a British survey of the Mediterranean, yet it was the island of Crete that most captivated his attention; his travels there became the subject of this two-volume account, published in 1837. The following year, Pashley's notes, collected artefacts and books were destroyed in a fire, so this work is all that remains from his expedition to the island. Crete at various points in its history had been ruled by Romans, Byzantines, Venetians and Ottomans. At the time of Pashley's arrival it was under Egyptian administration and there were palpable tensions between Christians and Muslims. In Volume 2, Pashley ends his travels in the south-western Samaria region. This volume also includes an appendix of Venetian manuscripts, and economic and demographic statistics from 1834.

Cambridge University Press has long been a pioneer in the reissuing of out-of-print titles from its own backlist, producing digital reprints of books that are still sought after by scholars and students but could not be reprinted economically using traditional technology. The Cambridge Library Collection extends this activity to a wider range of books which are still of importance to researchers and professionals, either for the source material they contain, or as landmarks in the history of their academic discipline.

Drawing from the world-renowned collections in the Cambridge University Library and other partner libraries, and guided by the advice of experts in each subject area, Cambridge University Press is using state-of-the-art scanning machines in its own Printing House to capture the content of each book selected for inclusion. The files are processed to give a consistently clear, crisp image, and the books finished to the high quality standard for which the Press is recognised around the world. The latest print-on-demand technology ensures that the books will remain available indefinitely, and that orders for single or multiple copies can quickly be supplied.

The Cambridge Library Collection brings back to life books of enduring scholarly value (including out-of-copyright works originally issued by other publishers) across a wide range of disciplines in the humanities and social sciences and in science and technology.

Travels in Crete

VOLUME 2

ROBERT PASHLEY

CAMBRIDGE
UNIVERSITY PRESS

CAMBRIDGE UNIVERSITY PRESS

Cambridge, New York, Melbourne, Madrid, Cape Town,
Singapore, São Paolo, Delhi, Mexico City

Published in the United States of America by Cambridge University Press, New York

www.cambridge.org
Information on this title: www.cambridge.org/9781108050838

This edition first published 1837
This digitally printed version 2012

ISBN 978-1-108-05083-8 Paperback

H. Schranz delᵗ.

L. Haghe Lith.

LUTRO

TRAVELS IN CRETE

BY

ROBERT PASHLEY ESQ.

FELLOW OF TRINITY COLLEGE CAMBRIDGE

VOLUME II

CAMBRIDGE
PRINTED AT THE PITT PRESS BY JOHN W. PARKER
PRINTER TO THE UNIVERSITY

LONDON
JOHN MURRAY ALBEMARLE STREET

M DCCC XXX VII

CONTENTS

OF

VOLUME II.

CHAPTER XXI.

CHAPTER XXII.

CHAPTER XXIII.

CHAPTER XXIV.

CHAPTER XXV.

CHAPTER XXVI.

CHAPTER XXVII.

CHAPTER XXVIII.

CHAPTER XXIX.

CHAPTER XXX.

CHAPTER XXXI.

CHAPTER XXXII.

CHAPTER XXXIII.

CHAPTER XXXIV.

CHAPTER XXXV.

CHAPTER XXXVI.

CHAPTER XXXVII.

CHAPTER XXXVIII.

CHAPTER XXXIX.

LIST OF DRAWINGS ON STONE

IN VOLUME II.

LIST OF ENGRAVINGS ON WOOD

IN VOLUME II.

[1] The height of the portion engraved is twelve feet two inches.

TRAVELS IN CRETE.

CHAPTER XXI.

STATUE FOUND AT THE SITE OF APTERA. SARCOPHAGUS FOUND AT ARVI. A LAMP EXHIBITING ZEUS AND HIS EAGLE.

THE engraving here presented to my reader represents the work of ancient art found at Aptera, and spoken of in the previous volume[1].

[1] Vol. I. p. 42. The statue has arrived in England, and is now in my possession.

While I staid at Khaniá, in April, I examined and had put together the fragments of the sarcophagus found at Arví, and thus discovered that several considerable gaps still existed in the monument. I thought it worth while to endeavour to render it as perfect as possible, and, before I left the island, I therefore caused fresh excavations to be made, on the distant spot where it was found, and had, ultimately, the great satisfaction of obtaining five additional fragments. When they were combined with those previously obtained, the sarcophagus was rendered almost as perfect as when it came from the chisel of the artist[2].

It is certainly interesting to see an ancient monument of Cretan sculpture. Scyllis and Dipoenos, who are spoken of as sons or disciples of Daedalos[3], were natives of Crete[4], and were among the most celebrated of ancient artists. They flourished about the fiftieth Olympiad, and, in their hands, the art of sculpture made great advances towards that wonderful perfection which it attained in ancient Greece[5].

The subject represented on this sarcophagus is interesting, though not of uncommon occurrence on similar monuments. The naked Bacchante on the left[6], whose hair flows unconfined down her back[7], is playing on a tympanum, an instrument common to the rites of both

[2] The pieces, on their arrival in England, were joined, under the inspection of Sir Francis Chantrey, and the sarcophagus was given, by Sir Pulteney Malcolm, to the University of Cambridge. It will be placed in the Fitz-William Museum.

[3] PAUSANIAS, II. p. 143.

[4] PAUSANIAS, l. c. CLEMENS ALEX. Protrept. p. 31. Σκύλης καὶ Δίποινος, Κρητικὼ οἶμαι ἀνδριαντοποιώ.

[5] PLINY, XXXVI. 4. "Marmore scalpendo primi omnium inclaruerunt Dipoenus et Scylis, geniti in Creta insula, etiamnum Medis imperantibus, priusque quam Cyrus in Persis regnare inciperet." WINCKELMANN, Werke, Vol. VIII. p. 309. QUATREMERE DE QUINCY, Le Jupiter Olympien, p. 180. foll. MUELLER, Handbuch der Archaeologie der Kunst, §. 82. HIRT, Geschichte der bildenden Kuenste bei den Alten, pp. 77-78.

[6] See the engraving below, opposite to p. 7.

[7] OVID, Ars. Am. I. 541.

Ecce Mimallonides sparsis in terga capillis.

Dionysos and Rhea[8], and said by Euripides to have been an invention of the Corybantes[9]. It was made of an animal's skin stretched on a hoop, and is frequently seen in the hands of these attendants on Dionysos[10]. Like the cymbal it was unknown to Homer's age[11], when the usage, even of that earlier invention the flute, was confined to the Phrygians[12], to whom its discovery is usually assigned[13], and who are said to have first employed it in the celebration of mystic rites[14].

On the body of the chariot is seen a panther, and the same animal is again observed to the right beyond the elephant, and also on the cornices at both ends of the sarcophagus. The presence of lions, tigers and panthers, in these processions of Dionysos, is usually accounted for by their supposed fondness for wine, which is both described by ancient authors[15], and implied in innumerable representations on works of ancient

[8] EURIPIDES, Bacch. 58.

Τύμπανα 'Ρέας τε μητρὸς ἐμά θ' εὑρήματα.

HORACE, Od. I. 18. 13.

[9] EURIPIDES, Bacch. 124.

Βυρσότονον κύκλωμα τόδε
μοι Κορύβαντες εὗρον.

See LOBECK, Aglaophamus, p. 1144. The Indians attributed the invention to Dionysos: DIODORUS SICULUS, III. 58. The Phrygians assigned it to Rhea: DIODORUS SICULUS, II. 38.

[10] MONTFAUCON, Antiquité expliquée, Tom. I. Par. II. Pl. CLXIII. 3. LE PITTURE D'ERCOLANO, Tom. I. Tav. XXI. p. 112. Tom. II. Tav. XXIX. p. 175. Tom. V. Tav. XXXV. p. 153. MUSEO CAPITOLINO, Tom. IV. Tav. XLVII.

[11] MILLIN, Monumens inédits, Tom. I. p. 164.

[12] OLYMPIODORUS, ap. Wyttenb. in Phaed. p. 146. et ap. Lobeck, Aglaoph. p. 298.

[13] Marsyas is said to have been its inventor: see LE PITTURE D'ERCOLANO, l. c. TELESTES in Athenaeus, XIV. p. 617.

[14] OXFORD MARBLES, p. 21. HIMERIUS, Ecl. XIII. p. 210. LUCRETIUS, II. 620. the notes on LE PITTURE D'ERCOLANO, Tom. II. Tav. XX. and XXI. HOECK, Kreta, Vol. I. p. 223. foll. LOBECK, Aglaophamus, Vol. I. p. 298.

[15] OPPIAN, Cyneg. III. 79. IV. 231 foll. STATIUS, Theb. IV. 658. MARTIAL, XIV. 107.

Nos Satyri, nos Bacchus amat, nos ebria tigris,
Perfusos domini lambere docta pedes.

art[16]. Thus the tiger was appropriately yoked to the car of Dionysos on his Indian expedition[17].

The youthful Deity is as usual standing in his car[18]: the weapon in his left hand is that with which he is generally armed[19]. In a well-known representation of him made by an ancient artist, Polycletus, he held this thyrsus in one hand and a drinking-cup in the other[20], and is thus doubly armed in several existing ancient monuments[21]. The thyrsus was a spear, the point of which was covered, sometimes with ivy[22], but more commonly, as in the present instance, by the cone of a pine: the shining steel was rarely allowed to meet the eye[23]. *Mitrae* are attached to the thyrsus, as they are to those of most other ancient monuments[24]: they were seen on that carried by the colossal figure of Nysa in Ptolemy's celebrated procession[25].

Unshorn locks, like those which we see fall over the shoulders of Dionysos, are generally found both in

16 GEMMAE MUSEI FLORENTINI, Tom. I. Tab. LXXXXII. PIERRES GRAVEES DU CABINET D'ORLEANS, Tom. I. Pl. 67. STUART'S ATHENS, Vol. I. Pl. XXX. (on the Choragic monument of Lysicrates.) MUSEO CHIARAMONTI, Tav. XXVIII. MUSEO BORBONICO, Tom. VII. Tav. LXII. and Quaranta, p. 6.

17 NONNUS, VIII. 36. MARTIAL, XV. 20.

18 An exception to the rule is found in the MUSEO PIO-CLEMENTINO, Tom. V. Tav. VII. where see VISCONTI, p. 13.

19 I need hardly quote the "Parce, Liber, Parce, gravi metuende thyrso" of HORACE.

20 PAUSANIAS, VIII. p. 665. SIDONIUS APOLLINARIS, XXII. 31.
Cantharus et thyrsus dextra laevaque feruntur.

21 See BUONARROTI, Osservazioni sopra alcuni medaglioni, p. 433. GEMMAE MUSEI FLORENTINI, LE PITTURE D'ERCOLANO, Tom. II. Tav. XIII. XVIII. Tom. III. Tav. II.

22 MACROBIUS, Saturnal. I. 19. "Cum thyrsum tenet, quid aliud quam latens telum gerit, cujus mucro hedera lambenti protegitur."

23 As in LE PITTURE D'ERCOLANO just quoted.

24 See BEGER's Thesaurus Brandenburgicus, p. 14. LE PITTURE D'ERCOLANO, Tom. III. p. 9. and VISCONTI, Museo Pio-Clementino, Tom. V. Tav. X. p. 19.

25 ATHENAEUS, V. p. 198. e. Μετὰ δὲ ταύτας ἤγετο τετράκυκλος—ἐφ' ἧς ἄγαλμα Νύσης—εἶχε δὲ ἐν τῇ ἀριστερᾷ θύρσον ἐστεμμένον μίτραις.

descriptions[26] and representations[27] of him: as are also the feminine features with which he is here represented[28]. On some monuments[29] he is even clad in woman's apparel:

[26] PINDAR, Isthm. VII. 3.

Ἦ ῥα χαλκοκρότου πάρεδρον
Δαμάτερος ἀνίκ' εὐρυχαίταν
ἄντειλας Διόνυσον.

TIBULLUS, I. Eleg. IV. 27.

Solis aeterna est Baccho Phoeboque juventus;
Nam decet intonsus crinis utrumque deum.

OVID, Met. III. 421.

Et dignos Baccho, dignos et Apolline crines.

Met. IV. princip.

Indetonsusque Thyoneus.

SENECA, l. c.

Spargere effusos sine lege crines.

MARTIAL, IV. Epigr. 45. In Lucian's Dialogues of the Gods, (Tom. I. p. 207. ed. Hemst.) Eros says to Zeus, ὡς ἥδιστον ποίει σεαυτόν, ἑκατέρωθε καθειμένος βοστρύχους—καὶ ὄψει ὅτι πλείους ἀκολουθήσουσί σοι τῶν Διονύσου Μαινάδων.

[27] See the MUSEO PIO-CLEMENTINO, Tom. II. Tav. XXVIII. and Visconti's observations, p. 56. ZOEGA, Abhandlungen, p. 24.

[28] MONTFAUCON, Antiquité expliquée, Tom. I. Par. II. Pl. CLIII. LE PITTURE D'ERCOLANO, Tom. II. Tav. XVIII. p. 117. WINCKELMANN, Monumenti antichi inediti, Tratt. prelim. p. XLI. (Werke, Vol. VII. p. 85.) WOBURN ABBEY MARBLES, Pl. XVII. MUSEO BORBONICO, Tom. VIII. Tav. LI. and Sir WILLIAM GELL's Pompeiana. He is called θηλύμορφος by EURIPIDES, Bacch. 353. OVID, Met. III. 607.

Virginea puerum ducit per litora forma.

MET. IV. 117.

Tibi enim inconsumpta juventus,
Tu puer aeternus, tu formosissimus alto
Conspiceris coelo: tibi quum sine cornibus adstas
Virgineum caput est.

The observations of VISCONTI (Museo Pio-Clementino, Tom. VI. Tav. VI.) serve as a commentary on the expression *quum sine cornibus adstas.* The CARMINA PRIAPEA, XXXVI.

Phoebus comosus, Hercules lacertosus,
Trahitque Bacchus virginis tener formam.

ALBRICUS PHILOSOPHUS, c. XIX. (in the Mythographi Latini of Van Staveren, p. 926.) "Erat enim imago sua facie muliebri."

[29] Dionysos is thus represented, clothed in a female dress, on the reverse of a Greek medal of Commodus, struck at Nicaea: see the NUMMOPHYLACIUM REGINAE CHRISTINAE, Tab. XLII. Num. 25. where Havercamp justly observes (p. 449.) that the God is represented "novo ritu, sedens videlicet, muliebri indutus veste longa, sceptrum laeva tenens, atque cantharum effundens in pantheram quae ante pedes est. Sine dubio ejusmodi signum Nicaeae exstitit." See also MAFFEI, Gemme antiche, Par. III. Tav. XXVII. pp. 44–5. GEMMAE MUSEI FLORENTINI, Tom. I. p. 162. MUSEO PIO-CLEMENTINO, Tom. VII. Tav. II. SENECA, Oed. T. 419.

Creveras

and Apollodorus says of him that he wore a girl's dress from the time when he was a child[30]; and also mentions his having received from Cybele a woman's tunic, in which he set out on his expedition to conquer India[31].

Hence, when ancient polytheism received its modern form, in Rome, the statue of the youthful Bacchus was easily converted into that of a female saint[32]: and hence also we obtain, I think, a sufficient explanation of the reason why the effeminate deity was represented clothed in Venus's pallá, in an ancient group which is praised by Pliny[33].

This effeminacy of Dionysos's dress and character was undoubtedly exhibited on the Attic stage by Aeschylus, in his Edoni, a passage of which play is preserved in the parody of Aristophanes[34].

Creveras falsos imitatus artus,
Crine flaventi simulata virgo,
Luteam pallam retinente zona.

30 APOLLODORUS, III. 11. 3.

31 APOLLODORUS, III. 5. 1. NONNUS also, XIV. 159. foll. represents Dionysos in a woman's dress.

Πῇ δὲ γυναικείην φορέων ψευδήμονα μορφήν
μιμηλὴ κροκόπεπλος ἐν εἵμασι φαίνετο κούρη
ἀρτιθαλής· φθονερῆς δὲ παραπλάζων νόον Ἥρης,
χείλεσιν ἀντιτύποισιν ἀνήρυγε θῆλυν ἰωήν,
καὶ πλοκάμοις εὔοδμον ἐπεσφήκωσε καλύπτρην,
θήλεα πέπλα φέρων πολυδαίδαλα· μεσσατίῳ δέ
στήθεϊ δεσμὸν ἔβαλλε, καὶ ὄρθιον ἄντυγα μαζοῦ
παρθενίῳ ζωστῆρι, καί, οἷά περ ἅμμα κορείης,
πορφυρέην λαγόνεσσι συνήρμοσε κυκλάδα μίτρην.

32 MIDDLETON, Letter from Rome, p. 160. Similar conversions are alluded to by DYER, in his poem, "The Ruins of Rome."

33 PLINY, N. H. XXVI. 4. "Satyri quatuor: ex quibus unus Liberum patrem palla velatum Veneris praefert." I wonder that VISCONTI (Museo Pio-Clementino, Tom. III. Tav. XL. p. 51.) should compare this statue, clothed in the palla of Venus, with that of the bearded Bacchus in his syrma, and should not, seemingly, have supposed the little wearer of Venus's palla to have had a feminine countenance.

34 ARISTOPHANES, Thesm. 136. Ποδαπὸς ὁ γύννις; from the Scholiast on which passage it is plain, that the words were addressed to Dionysos, in the Edoni, and ought undoubtedly to have a place among the fragments of that play, (where it is accordingly given to them by Mr DINDORF, Poet. Scen. Graec. Fr. Aesch. 55. p. 6) although we vainly look for them in the volume of the Reverend Professor Scholefield, the last English editor of Aeschylus, (see Aeschyl. Scholef. ed. 2da Cantab. 1830. pp. 386-7.) Γύννις is explained

SARCOPHAGUS FOUND AT ARVI.

It seems, even from what little has been stated, that both poets and artists did much towards rendering the sex of Dionysos ambiguous. Later authorities[35] go still further, and assert that he was actually both male and female.

Hence the epithet *androgynous* is bestowed on a statue of Dionysos, by the Christian author Theodoret[36], in a passage of his Ecclesiastical History, where he is speaking of the persecution suffered by the Church when the Emperor Julian was re-establishing the old religion of the Roman world.

The youth on whose shoulder the right arm of Dionysos reposes must next be considered. We find in the Dionysiacs of Nonnus a youthful companion of the God whose society he preferred even to the feasts of the Gods and the friendship of Zeus[37], and of whose beauty the poet gives an elaborate and glowing description[38]. His name was Ampelos, and I have no doubt that he is meant to be represented by the youth in the car. The little tail seen on his back, and his pointed

explained by HESYCHIUS and SUIDAS, and is used by PHILOSTRATUS, Vit. Sophist. II. 31. 2. where Perizonius supposes it to indicate Heliogabalus.

[35] This is the account given in the ORPHIC POEMS: Hymn XXX. *Κικλήσκω Διόνυσον—πρωτόγονον διφυῆ.* Hymn XLII. 4.

Ἄρσενα καὶ θῆλυν διφυῆ λύσειον Ἴακχον.

ARISTIDES, p. 52. 8. *Ταῦτ' ἄρα καὶ ἄῤῥην τε καὶ θῆλυς ὁ θεός.* Again, p. 52. 10. *Ἐστὶ δὲ τῇ φύσει καὶ τὴν μορφὴν προσεοικώς· ὥσπερ γὰρ δίδυμος πάντη αὐτὸς πρὸς ἑαυτόν ἐστι.* See VOSS, Mythologische Briefe, and, on the other hand, HEINRICH's academical dissertation, "Hermaphroditorum, artis antiquae operibus illustrium, origines et causae."

[36] His words are, (Eccles. Hist. III. 7.) *Ἐν Ἐμέσῃ δὲ τῇ ὁμόρῳ πόλει, Διονύσῳ τῷ γύννιδι τὴν νεόδμητον ἀφιέρωσαν ἐκκλησίαν, τὸ καταγέλαστον καὶ ἀνδρόγυνον ἐν αὐτῇ ἱδρύσαντες ἄγαλμα.*

[37] NONNUS, X. 284.

Οὐ πόθον ἕλκω
νέκταρος, ἀμβροσίης δ' οὐ δεύομαι· οὐκ ἀλεγίζω,
Ἄμπελος εἰ φιλέει με, καὶ ἐχθαίρει με Κρονίων.

OVID, Fasti, III. 409.

Ampelon intonsum, Satyro Nymphaque creatum,
Fertur in Ismariis Bacchus amasse jugis.

[38] NONNUS, l. c. v. 175—192.

goat-like ears, sufficiently identify the figure with that of the young Satyr[39] whom Dionysos so fondly loved.

The attitude of Dionysos and Ampelos in the car is one in which they are found on several other ancient monuments. In a well-known group, Eros has conducted Dionysos to the place where Ariadne is sleeping, and the God's left arm rests on the shoulder of his youthful companion[40]. Among the statues of what was once the Ducal palace at Venice, is a group consisting of Dionysos and Ampelos[41]: they are also seen on a sarcophagus in the Vatican[42]. In other monuments Dionysos's right hand rests on the same youth[43]: and we find them together in a bronze group discovered lately at Pompeii, and now in the Studij at Naples[44]. A young Satyr supports the left arm of the bearded Dionysos in an ancient monument of Greek sculpture contained in the noble collection of the Vatican[45], and

[39] NONNUS, X. 208.

Εἰ Κρονίδης μὲ φύτευσε, σὺ δὲ χθονίης ἀπὸ φύτλης
βουκεράων Σατύρων μινυώριον αἷμα κομίζεις,
ἶσον ἐμοὶ βασίλευε θεῷ βροτός.

Compare WINCKELMANN, Monumenti antich. ined. Par. prim. p. 6. On the tails of the Satyrs see LI BRONZI D'ERCOLANO, Tom. II. Tav. XXXVIII. XXXIX. and p. 146. If the presence of the tail is indispensable in the figure of Ampelos, as I think it must be considered, Mr Becker's conjecture respecting the Dresden statue (AUGUSTEUM DRESDENS, Vol. III. p. 90.) cannot be received. Visconti's assertion, too, (MUSEO PIO-CLEMENTINO, Tom. IV. Tav. XXII. p. 48.) that on a lion's back "siede senza freno il fanciullo Ampelo," seems difficult to admit; for, fitly as Ampelos is placed on a lion's back, (see NONNUS, XI. 66. foll.) yet we can hardly venture to call by his name a figure in which all the marks of a caprine origin are wanting.

[40] Published elsewhere, and in Mr HIRT's Bilderbuch, Tab. X. Fig. 5.

[41] STATUE DI SAN MARCO, Tom. II. Tav. XIV. XXVI. where the God's arm rests on the young Satyr's shoulder. See also the STATUAE MUSEI FLORENTINI, Tab. XLVIII.

[42] MUSEO PIO-CLEMENTINO, Tom. V. Tav. VIII.

[43] In the MUSEO PIO-CLEMENTINO, Tom. I. Tav. XLII. "Bacco con Fauno:" and Sir WILLIAM GELL's Pompeiana, Vol. II. Pl. LXXVIII.

[44] MUSEO BORBONICO, Tom. III. Tav. IX. Bacco ed Ampelo. Gruppo di bronzo alto palme tre oncie due. Ampelos alone appears in the MUSEUM FLORENTINUM, Gemmae, Tom. I. Tav. LXXXVIII. 8.

[45] MUSEO PIO-CLEMENTINO, Tom. IV. Tav. XXV. VISCONTI, p. 52. observes, "Un fauno per molle commodità il sostiene sotto il sinistro cubito."

the attitudes of the two figures remind us of those in which we have seen the youthful God and Ampelos. Elsewhere also the thyrsus-bearing and bearded God is accompanied by Ampelos[46]. In different museums other groups are found in which figures have been taken for Ampelos, sometimes, perhaps, without sufficient authority[47].

He is sometimes called the genius of Dionysos, and seems to be entitled to rank as one of those ministering powers who attended on and performed the behests of superior deities. Acratos was another genius of the same God: he is mentioned by Pausanias[48], is called Acratopotes in Athenaeus[49], and is represented, on ancient monuments, as a winged boy, with a smiling, and, not unfrequently, a drunken countenance[50].

We know the names of very few of the thirty thousand demons, spirits of departed heroes, who, according to Hesiod[51], exist on the earth, and superintend the

[46] D'HANCARVILLE, Antiquités Etrusques Grecques et Romaines, Tom. II. Pl. V. pp. 110–111. STATUAE MUSEI FLORENTINI, Tom. V. Tab. XLVIII.

[47] For instance, in the "Bacchus cum Ampelo" of the MUSEUM FLORENTINUM, Tom. III. Tab. XLVII. p. 52. the supposed Ampelos bears no mark of his descent Σατύρων ἀπὸ φύτλης. The same thing may be said of the figure taken for Ampelos in the MUSEO CAPITOLINO, Tom. IV. p. 249. WINCKELMANN, however, takes the former figure for a Satyr: Werke, Vol. III. p. x. See Obss. p. 4. Again, in a group of the MUSEO BORBONICO, Tom. VIII. Tav. LI. "Bacco da l'uva alla vite," I should hardly take the little boy seen receiving a bunch of grapes, for Ampelos, who is, however, in all probability meant to be represented by the figure of a youth drawn in a car, on a bas-relief of the Vatican, MUSEO PIO-CLEMENTINO, Tom. IV. Tav. XXV. Visconti's supposition, that it is Acratos, is clearly indefensible, if the view which I have taken of these two genii is correct. In another group, MUSEO PIO-CLEMENTINO, Tom. IV. Tav. XX. Visconti describes Bacchus as "retto dal giovinetto Acrato o Ampelo;" where the sculptor must have designed to represent Ampelos alone.

[48] PAUSANIAS, I. p. 2. Δαίμων τῶν ἀμφὶ Διόνυσον Ἄκρατος.

[49] ATHENAEUS, II. p. 39.

[50] GORII Gemmae Musei Florentini, p. 95. MUSEO PIO-CLEMENTINO, MUSEO BORBONICO, Tom. V. Tav. VIII. Tom. VII. Tav. LXII.

[51] HESIOD, Works and Days, I. 250.

Τρὶς γὰρ μύριοι εἰσὶν ἐπὶ χθονὶ πουλυβοτείρῃ
ἀθάνατοι Ζηνὸς φύλακες θνητῶν ἀνθρώπων.

affairs of men. Those of Jupiter and Bacchus are said by Winckelmann to be the most common[52]. Ancient authors mention the names of four such attendants on Aphrodite alone. They were Tychon, Orthages, Gigon, and Genetyllis[53]. Demeter had two[54], Adreus and Calligeneia. Artemis likewise had a demon or genius, Genetyllis[55].

This seems also to have been the relation of the Corybantes, Curetes, Dactyli and Cabiri to the Gods[56]. The Curetes are called by an unknown poet in the Eclogues of Stobaeus, "assessors of Rhea the mother of Zeus[57]," and these demons seem all to have been regarded as powerful superintendents of mankind[58], as secondary controllers of fate[59], and as mediators between God and man[60], resembling, in no slight degree, the saints of the Roman Catholic mythology. They were called genii by the Romans, and were supposed to watch over individuals. Each Emperor had his guardian

[52] WINCKELMANN, Monumenti Antichi inediti, p. 6.

[53] See PAUSANIAS, I. p. 2. with passages of CYRIL, and HESYCHIUS, quoted by LOBECK, Aglaoph. p. 1235. and the SCHOLIAST ON ARISTOPHANES, Thesm. 130.

[54] See the GREAT ETYMOLOGICON, and the SCHOLIAST ON ARISTOPHANES, Thesm. 299.

[55] BENTLEY, on Horace, Carm. Saec. v. 16. where he reads, "Sive tu Lucina probas vocari, Seu Genetyllis."

[56] STRABO, X. p. 472.

[57] STOBAEUS, Ecl. I. p. 68.

Κουρήτων οἳ ματρὶ Διὸς Ῥέῃ ἐντὶ πάρεδροι.

[58] The stoics used to call them *ἐπόπτας ἀνθρωπίνων πραγμάτων*, DIOGENES LAERTIUS, VII. 15.

[59] ALCIPHRO, I. Ep. 20. *Ἀλλ' ὦ μοιραῖοι θεοὶ καὶ μοιραγέται δαίμονες.*

[60] PLATO, Sympos. p. 202. e. *Πᾶν τὸ δαιμόνιον μεταξύ ἐστι θεοῦ τε καὶ θνητοῦ, διαπορθμεῦον θεοῖς τὰ παρ' ἀνθρώπων καὶ ἀνθρώποις τὰ παρὰ θεῶν*, where VON AST's note may be consulted. PLUTARCH, de Isid. et Os. p. 351. VAN DALE, De Orig. et Prog. Idolol. Diss. c. III. Further information respecting these Greek demons may be obtained from that storehouse of profound learning and acute criticism, Professor LOBECK's Aglaophamus, p. 1233. foll. as well as from the writings of several of our own countrymen, who have considered the question respecting these profane spirits in relation to sacred demonology.

genius[61], and, even among the Greeks, traces of the same belief may be found[62]. The Roman genii were also supposed to preside over particular places[63]: but we find no traces of any such notion attached to their demons by the Greeks[64].

The car is drawn by a Centaur and Centauress. These monsters were supposed by the naturalist Pliny really to have existed[65], but the poet Lucretius was less credulous[66]. They are described, even in the Odyssey[67], as fond of wine: and it is said that the smell of it, when it was produced by Pholus in honour of his guest Hercules, immediately drew them to his house, and thus led to the contest between them and the son of Alcmena[68]. Nonnus describes the Centaur

61 GRUTER and MURATORI both afford sufficient proof of this. The Genius of Julian seems to be exhibited in the MUSEUM FLORENTINUM, Tom. I. p. 49. and Tab. XVIII. WINCKELMANN, Monumenti Ant. ined. p. 6. quotes the verse of Martial, VI. Ep. LX. 10. "Victurus Genium debet habere liber."

62 PLATO assigned one to every individual: see the Phaedon, p. 107. d. 'Ο ἑκάστου δαίμων ὅσπερ ζῶντα εἰλήχει. p. 113. d. Oἶ ὁ δαίμων ἕκαστον κομίζει. MUSEO CAPITOLINO, Tom. IV. Tav. XXV. p. 135.

63 Many inscriptions are found in Gruter, and in other collections, addressed "*Genio loci.*" See the PITTURE D'ERCOLANO, Tom. I. Tav. XXXVIII. SYMMACHUS, X. Ep. LXI. p. 442. "Ut animae nascuntur ita populis natales Genii dividuntur." PRUDENTIUS, II. in Symmach. 71. foll. LOBECK, Aglaoph. pp. 595–6. Thus in modern times each town of Italy has its patron Saint, who, I think, is to be compared with the local genius of the old Italian superstition, and not with the tutelary *deities* of *Greek* cities.

64 See BURMANN, on Rutil. Tom. II. p. 233. and LOBECK, l. c. The Τυχεῖα of Greek cities, temples of these local Genii, were built after the spread of the Roman Empire. VALESIUS, on Socrat. Hist. Eccles. III. 11. p. 187. on Euseb. VIII. 11. p. 433. and the BRONZI D'ERCOLANO, Tom. II. p. 107.

65 PLINY, N. H. VII. 3.

66 LUCRETIUS, V. 19.

67 ODYSSEY, XXI. 295.

Οἶνος καὶ Κένταυρον ἀγακλυτὸν Εὐρυτίωνα
ἄασ' ἐνὶ μεγάρῳ μεγαθύμου Πειριθόοιο.

OVID, Met. XII. 220. VIRGIL, Georg. II. 455. STATIUS, Theb. V. 261. DIO CHRYSOSTOM, Orat. XVII.

68 APOLLODORUS, II. 5. 4. on which passage Heyne quotes TZETZES, Chil. V. 120. and STESICHORUS, in Athenaeus, XI. p. 449. b. See also POLYAENUS, Strateg. I. c. III. 1.

as voluntarily yoking himself to the car of Dionysos[69], and we see, from the emptied drinking-vessel of the Centauress before us, that the God did not neglect to requite them for their zeal in his service. This two-handled vessel is the cantharus which was peculiarly appropriated to Dionysos[70].

Centaurs are exhibited to us in many works of ancient art, and the car of Dionysos is frequently drawn by them[71]. Their place is, however, sometimes taken by lions, tigers or panthers, and, on one single monument, by horses[72]. Centauresses are not so common as Centaurs, and are never represented on the sculpture of the temples of Greece. When met with they are mostly in attendance on Dionysos, as in the work before us[73]. The branch of pine borne by the Centauress reminds me of a dendrophorous Centaur on a fictile vase engraved in Sir William Hamilton's collection[74]. I may also refer to the dendrophori in Ptolemy's Bacchic procession[75], to the Bacchanti in Palaephatus[76], and to

69 NONNUS, XIV. 265. He first describes the exertions of the Corybantes in harnessing the lions and panthers, *χεῖλος ἐπισφίγξαντες ἀπειλητῆρι χαλινῷ*, and then speaks of the Centaurs:

> *Καὶ λασίην Κένταυρος ἔχων φρίσσουσαν ὑπήνην*
> *εἰς ζυγὸν αὐτοκέλευστος ἑκούσιον αὐχένα τεῖνεν,*
> *καὶ Σατύρων πολὺ μᾶλλον ἔχων πόθον ἥδεος οἴνου,*
> *ἡμιτελὴς χρεμέτιζεν ἀνήρ, κεκερασμένος ἵππῳ·*
> *ἱέμενος Διόνυσον ἑοῖς ὤμοισιν ἀείρειν.*

70 MACROBIUS, Saturn. v. 21. "Scyphus Herculis poculum est, ita ut Liberi patris cantharus."

71 See BUONARROTI, Osservazioni sopra alcuni Medaglioni, p. 428. fol. SPANHEIM, de praestantia et usu numism. antiq. Tom. II. p. 16. BANIER, Histoire de l'Academie, Tom. III. p. 19. MONTFAUCON, Antiquité Expliquée, Tom. I. Part. I. Pl. CLV. CLVI. MUSEO CAPITOLINO, Tom. IV. Tav. XXXII. et XXXIII. LE PITTURE D'ERCOLANO, Tom. I. Tav. XXVI. and XXVIII. VISCONTI, Museo Pio-Clementino, Tom. I. Tav. XLII. Tom. IV. Tav. XXII. XXVI. Tom. V. Tav. VII.

72 See VISCONTI, Museo Pio-Clementino, Tom. IV. Tav. XXIV. p. 49.

73 On these centauresses see WINCKELMANN, Monumenti inediti, p. 107. and VISCONTI, M. P. C. Tom. IV. Tav. XXI.

74 COLLECTION OF ETRUSCAN, GREEK AND ROMAN ANTIQUITIES, Vol. I. Pl. XLII.

75 ATHENAEUS, V. l. c.

76 PALAEPHATUS, c. 34. 3.

Visconti's observations respecting a similar figure on a bas-relief of the Museo Pio-Clementino[77].

We now come to the figure with cloven hoofs, a goat's legs and a long beard, whose boisterous drunken mirth seems to draw a reproof from the venerable person by whom he is addressed. Those who still follow the common opinion of the archaeologists of the last century[78], and of some even of the present day, and confound the Satyrs of the Greeks with the Fauns of the Romans, may suppose a Satyr to be represented by this figure. But it seems clear that the Greek Satyrs are always exhibited with the feet of men[79], differing only from the Sileni in being young, while the latter are of a more advanced age[80]. The Fauns of the Romans had the extremities of a goat, and derived their origin from Faunus, a rural God corresponding with the Greek Pan[81].

I suppose the figure on this Greek sarcophagus to represent Pan himself, and it answers very well to Lucian's description in the dialogue between him and his father Hermes[82]. The God addresses him, "You that have horns, and such a nose, and a matted beard, and cloven feet, and a tail dangling behind you." The common story of Pan's parentage sufficiently accounts for the peculiarities of his form. He was said to have

77 VISCONTI, M. P. C. Tom. V. Tav. VII.

78 All from Montfaucon to Winckelmann and Visconti might be named.

79 See VOSS, Mythologische Briefe, Vol. II. LANZI, De' Vasi antichi dipinti, diss^{e}. 2da. §. VII. Also the AUGUSTEUM DRESDENS.

80 CHOEROBOSCUS, in Etymolog. Magn. *Σειληνοὶ λέγονται οἱ γέροντες τῶν Σατύρων.* PAUSANIAS, I. p. 54. *Τοὺς ἡλικίᾳ προέχοντας τῶν Σατύρων ὀνομάζουσι Σιληνούς.* NONNUS, XIV. 101. XXXIV. 140. PERIZONIUS, on Aelian, V. H. III. p. 245. LANZI, l. c. §. VI.

81 HORACE, Od. I. 17. 2. OVID, Fasti, II. 280. LANZI, Saggio di lingua Etrusca, Vol. II. and l. c. §. VII. and also BOCHART, Hierozoicon, Vol. III. p. 822. Thus the *panics* which the Greeks attributed to Pan (PAUSANIAS, X. p. 855. POLYAENUS, Strateg. I. 2. SUIDAS, v. *Πανικῷ δείματι*) were assigned by the Romans to Faunus. DIONYSIUS HALICAR. V. XVI. *Τούτῳ (Φαύνῳ) ἀνατιθέασι τῷ δαίμονι Ῥωμαῖοι τὰ Πανικά.*

82 LUCIAN, Dialogues of the Gods, XXII. (Tom. I. p. 269. ed. Hemst.) *Κέρατα ἔχων, καὶ ῥῖνα τοιαύτην, καὶ πώγωνα λάσιον, καὶ σκέλη δίχηλα καὶ τραγικά, καὶ οὐρὰν ὑπὲρ τὰς πυγάς.*

been the fruit of an amour between Penelope and Hermes, in which the God had assumed the form of a goat[83]; a legend which will remind the reader of the customs spoken of by the Jewish Lawgiver[84], and which are known from Pindar[85], Herodotus[86] and Plutarch[87], to have been observed in Egypt, as well as of some works of ancient art which have come down to us[88]. Pan is spoken of as the son of Hermes and Penelope by several authors[89]; but another tradition deprives him of his divine father, and, at the same time, affords an etymology of his name[90]. Pan and Silenos were Bacchus's generals on the Indian expedition, and the well-known aid which Pan rendered the Athenians at Marathon shews how valuable his military services were supposed to be. Now, however, he has laid aside his martial character, and appears, conformably with the mention of him by Pindar[91], Aeschylus[92] and other authors[93], as a dancing attendant on the God of wine.

[83] LUCIAN, l. c. p. 270. and Hemsterhusius's learned note.

[84] LEVITICUS, XVIII. 23. 27. and especially 30. Again, XX. 16.

[85] PINDAR, in Strabo, XVII. p. 802. *Ὅθι τράγοι γυναιξὶ μίσγονται.*

[86] HERODOTUS, II. 46. *Ἐν τῷ νομῷ τούτῳ—γυναικὶ τράγος ἐμίσγετο ἀναφανδόν.* JABLONSKI, Pantheon Aegyptiorum, II. VII.

[87] PLUTARCH, Gryllus, p. 989. a.

[88] Every one who has visited the lock-up class in the Studij at Naples will remember a masterpiece in its way. A singular tête-à-tête is represented in the Pierres Gravées du Cabinet d'Orleans, Tom. I. Pl. 75.

[89] HERODOTUS, III. 145. PLUTARCH, de orac. defect. Tom. II. p. 419. HYGINUS, Tab. CCXXIV. "Pan Mercurii et Penelopes filius." TZETZES, on Lycophron, 772. In the second Altar of DOSIADAS, Ulysses is spoken of as *Πανὸς ματρὸς εὐνέτας.*

[90] GREGORY NAZIANZEN, in Orat. III. Tom. I. p. 81. ed. Paris, 1609. *Καὶ τὸν τραγόπουν καὶ τὸν σεμνὸν Πᾶνα, τὸν ἐκ πάντων μνηστήρων ἕνα θεόν, καὶ λαβόντα τὴν ὕβριν ὥσπερ ἦν ἄξιον.* Compare the Commentary of ELIAS CRETENSIS, Tom. II. p. 337. NATALIS COMES, v. 6. p. 652. and MUNCKER, on Hyginus, l. c.

[91] PINDAR, Fr. 69. ed. Boeckh. *Χορευτὴς τελεώτατος θεῶν.*

[92] AESCHYLUS, Pers. 448. *Ὁ φιλόχορος Πάν.*

[93] ARISTIDES, Tom. I. p. 53. *Διδόασι δ' αὐτῷ (Διονύσῳ) καὶ τὸν Πᾶνα χορευτήν, τελεώτατον θεῶν ὄντα, ὡς Πίνδαρός τε ὑμνεῖ καὶ οἱ κατ' Αἴγυπτον ἱερεῖς κατέμαθον.* Pan is called *ὀρχηστὴς* in an old scolion: ATHENAEUS, XV. p. 694. d. *Ὦ Πάν, Ἀρκαδίας μέδων κλεεννᾶς, Ὀρχηστὰ Βρομίαις ὀπαδὲ Νύμφαις*: see ILGEN, Carm. Conviv. Graec. pp. 13–14.

Pan, the shortness of whose stature is generally to be remarked, occupies a similar situation just before the car of Dionysos, and following Silenos's group, on a Bacchanalian procession in the Duke of Bedford's gallery[94], and on several other ancient monuments.

The vessel which Pan has overturned is of unusual dimensions. The Cretan artist probably meant to represent the amphora most commonly seen in his own island, and which appears to have been of a very large size[95].

The clothed and bearded figure, whose left arm is thrown over the shoulder of a woman, while his right is stretched out towards Pan, with whom he seems to be expostulating, is evidently old Silenos, the preceptor of Dionysos, and his companion in the Indian expedition. Silenos had long before distinguished himself for his martial prowess in the combat of the Gods and Giants, in which he fought on the right hand of Dionysos, and slew Encelados[96].

Although the name Silenos is applied to all aged Satyrs[97], yet human ears serve to distinguish the companion of Dionysos from the rest of them[98], and he is frequently represented clothed[99], as is the case here, although undoubtedly he is more commonly naked on ancient monuments. He appears before us, on this sarcophagus, in his more dignified character, as the

[94] WOBURN ABBEY MARBLES, Plate XII.

[95] PHILOSTRATUS, Heroic. p. 34. ed. Boissonade. Τὸ γοῦν κράνιον, ἐμφορησάντων ἡμῶν εἰς αὐτὸ οἶνον, οὐδὲ ὑπὸ δυοῖν ἀμφορέοιν ἐνεπλήσθη, τῶν ἐκ Κρήτης.

[96] EURIPIDES, Cyclops, 6.

"Επειτα δ' ἀμφὶ γηγενῆ μάχην δορός
ἐνδέξιος σῷ ποδὶ παρασπιστὴς γεγώς,
'Εγκέλαδον, ἰτέαν ἐς μέσην θενών, δορί
ἔκτεινα.

[97] Above, p. 13.

[98] VISCONTI, Museo Chiaramonti, Tav. XLVI. and Museo Pio-Clementino, Tom. IV. Tav. XXVII. p. 56.

[99] WINCKELMANN, Pierres gravées de Stosch, 1470. MUSEO CHIARAMONTI, Tav. XXXIV. and VISCONTI's observations, p. 82.

preceptor of Dionysos[100], as a general, or as a philosopher[101]. This philosophical gravity seems, however, to be combined with drunkenness, as in Virgil's Eclogue. We must not forget the philosophical character of the old drunkard, when we find Socrates, whose personal appearance in some degree represented that of Silenos[102], compared to him by Alcibiades in the Symposion of Plato.

Silenos seems to be employed in defending the lady under his arm against the designs of Pan, whose character is too well-known to need any illustration[103].

As Visconti has sometimes confounded Acratos with Ampelos, so a living antiquary, Dr Giovanni Labus of Milan, has supposed[104] that Acratos and Silenos are different names of one and the same individual.

Of the elephant, emblematic of Dionysos's victorious return from the east, it is unnecessary to say any thing. The instrument, with which the driver, mounted on the animal's neck, directs its motions, is like that used in India, for the same purpose, at the present day.

The last figure, on the front of the sarcophagus, is that of a Satyr carrying a skin full of wine. Satyrs

[100] See BECKER's observations in the AUGUSTEUM DRESDENS, Vol. II. Pl. LXXI. He mentions the Borghese statue of Silenos holding the infant God in his arms, as an example of the first of the three classes in which he arranges ancient representations of him.

[101] See PERIZONIUS, on Aelian, V. H. III. 18. and the passages of DIODORUS SICULUS, CICERO, VIRGIL and PLUTARCH, in HEYNE's Argument to Virgil's sixth Eclogue, and in the PIERRES GRAVEES DU MUSEE D'ORLEANS, Tom. I. p. 252.

[102] See AST, on Plat. Sympos. §. XL. p. 216.

[103] THEODORET, when speaking of similar processions (Therap. VII. p. 885. Schulz. quoted by LOBECK, Aglaoph. p. 200.) says: Ἐν ταύταις ταῖς ὁμηγύρεσι πᾶν εἶδος ἀκολασίας ἐτολμᾶτο—τὰ παρὰ τῶν ὀργιαστῶν δρώμενα εἰς πᾶν ἀσελγείας τοὺς ὁρῶντας ἠρέθιζε—ὁποῖα τῶν Σατύρων ὁ χορὸς ἠσέλγησε δημοσίᾳ πομπεύων, ἔχων ἐν μέσῳ τὸν Σειληνὸν καὶ τὸν Πᾶνα, τὸν μὲν ἐπιλυττῶντα ταῖς Νύμφαις, τὸν δὲ λελυμένον ὑπὸ μέθης.

[104] LABUS, Museo della reale Accademia di Mantova, Vol. II. p. 180. So also ZOEGA, Abhandlungen, p. 26.

are often thus represented: sometimes they are mounted on their well filled wine-skins. Similar figures used to adorn fountains, and the water which issued, instead of wine, from their skins, is the subject of some epigrams in the Greek Anthology. Visconti explains a statue of Silenos, carrying a wine-skin, by referring to the military use of skin-bearers (utricularii) and to the service rendered, according to Nonnus, by such skins, to Dionysos on his expedition into India. A very early use of skins was for this excellent purpose of holding wine: when no longer so employed they still served for oil and water, as they do in Greece and Turkey even to the present day. The water-carriers of Khaniá and Megálo-Kástron, of Smyrna and Constantinople, have no other vessel wherein to hawk water about their cities; and the Cretan peasants are daily seen driving towards the principal ports mules, each of which has a great skin filled with oil suspended on either side of its pack-saddle.

The riotous companions of Dionysos, whom we see before us on the sarcophagus, have continued to exist under changed names, and with attributes somewhat altered, even since the general spread of Christianity. We find such points of resemblance between the names and attributes of these *kobali*[105], and those of the German *kobolds*[106], and the English *hob-goblins*[107], of the present day, that we cannot refrain from believing the demons of the old mythology still to exist in the popular superstitions of both England and Germany.

It has been conjectured, that the representation of these processions on tombs, may indicate that the dead person had been initiated in the Bacchic rites. The

105 HARPOCRATIO, Κόβαλοι, δαίμονες εἰσί τινες σκληροὶ περὶ τὸν Διόνυσον.

106 KEIGHTLEY, Fairy Mythology, Vol. II. pp. 41—68. ADELUNG, Woerterbuch, under Cobold. LOBECK, Aglaoph. p. 1320. J. GRIMM, Deutsche Mythologie, p. 286.

107 KEIGHTLEY, F. M. Vol. II. pp. 104 and 121.

festive scene exhibited on the cornice surmounting the part of the monument already described, may allude to the enjoyments supposed to be reserved, in the other world, for those who had been initiated in this. We see the lyre and the bowl, the usual accompaniments of celestial banquets, according to the poetry and religion of Greece; and which were naturally assigned to the blessed in the Pagan Elysium[108]. The sculptor has also added the rewards which Mohammed afterwards promised to the faithful, in the female figures seen among the revellers.

At one end of the sarcophagus we have Pan[109] exhibited, with two mischievous Loves, who are amusing themselves at his expense.

The other extremity of the monument represents a less common and much more interesting subject. The

[108] Those Κύκλιοι χοροί, καὶ μουσικὰ ἀκούσματα, ξυμπόσιά τε εὐμελῆ, καὶ εἰλαπίναι αὐτοχορήγητοι, καὶ ἀκήρατος ἀλυπία, καὶ ἡδεῖα δίαιτα, which are found εἰς τὸν τῶν εὐσεβῶν χῶρον, according to the author of the Axiochus, p. 164. ed. Fisch. Compare BOECKH, on Pindar, Tom. II.

[109] If any one should be disposed to regard this figure as representing Priapus, who was δυσειδὴς καὶ βαθυαίδοιος, he might quote, in defence of the goat's legs and the horns which we see before us, the words of PHURNUTUS, p. 204. Gale. Τάχα μὲν διὰ τοῦτο καὶ κεράστην καὶ δίχηλον πλάττοντες— ἴσως δ' ἂν οὗτος καὶ ὁ Πρίαπος εἴη.

youthful Dionysos is here carried in his cradle by two Satyrs.

The infancy of the Theban God was watched over by the Nymphs, or by the Muses, according to the testimony of many ancient authors; and, in one very spirited and beautiful monument of ancient art[110], the infant is swung in his basket or cradle, by a Nymph and a Satyr. Our artist has given the charge to two Satyrs alone. The torches borne by them remind us that the mystic rites of Dionysos used to be celebrated in the night-time.

Another monument of ancient art, made of more fragile materials than those which have been represented and described in this chapter, was obtained by me in the island. It is an earthenware lamp, on which are seen[111] Zeus and his eagle.

Many authors speak of the eagle, which carried off Ganymedes from his native fields, and conveyed him into

[110] WINCKELMANN, Monumenti antichi inediti, Parte Prima, Cap. XX. Also engraved in the Description of Ancient Terracottas in the British Museum, Pl. XXIV. n. 44.

[111] See the engraving at p. 21.

the presence of Zeus[112]; and some describe that deity as having himself assumed the eagle's form on the occasion[113]. There are also other traditions, which would equally account for the appearance of the king of birds in company with Zeus, both on Cretan coins, and on the lamp in question[114]. According to Homer, "the Gods" carried off Ganymedes on account of his beauty, and that he might pour out nectar for Zeus[115]. Echemenes, who wrote on Crete, mentions a tradition of this island, that it was Minos, and not Zeus, by whom the youth was carried off[116]; and the Chalcidians of Euboea also gave the credit of the exploit to the same mythical personage[117].

[112] See HEMSTERHUSIUS, on Lucian, Tom. I. p. 210. HEYNE, on Apollodor. p. 741.

[113] LUCIAN, l. c.

[114] See SERVIUS, on Virgil, Aen. I. 398. SPANHEIM, on Callimachus Hymn to Zeus, 68.

[115] HOMER, IL. XX. 234. SCHOLIAST, on Apollonius Rhodius, III. 115.

[116] ATHENAEUS, XIII. p. 601. e. *Ἐχεμένης γοῦν, ἐν τοῖς Κρητικοῖς, οὐ τὸν Δία φησὶν ἁρπάσαι τὸν Γανυμήδην, ἀλλὰ Μίνωα.* EUSTATHIUS, on IL. XX. p. 1205. HEYNE, on Apollodor. p. 532.

[117] HEMSTERHUSIUS, l. c. MEINEKE, on Euphorio, p. 10. SCHNEIDEWIN, on Ibycus, pp. 112—115.

CHAPTER XXII.

DEPARTURE FROM KHANIA. PLATANIA PROBABLY THE SITE OF PERGAMOS. RIVER IARDANOS. SUPPOSED SITE OF CYDONIA AT IERAMI. ARRIVAL AT THE MONASTERY OF GONIA. MONASTIC INSTITUTIONS. RELIGIOUS PAINTINGS OF THE GREEKS. SITE OF DICTYNNAEON.

April 21.

I LEFT Khaniá a little before twelve o'clock, and had scarcely passed the gate of the city before I heard the noon salute fired by the guns of the fortress, in honour of the lesser Bairám[1], which commenced yester-

[1] I have called this the lesser Bairám in conformity with the common practice both of the people in the country and of European writers who have spoken of the Mohammedan feasts, although no doubt, as has been observed by RELAND, de religione Mohammedica, p. 109. ed. alt. Traj. ad Rhen. 1717. (where he corrects D'HERBELOT, Bibliothèque Orientale, Art. Beirám,) and SALE, (Preliminary Discourse, §. 7. p. 200.) the feast succeeding the Ramazán is properly the lesser Bairám, and this is the greater. No one who has witnessed the celebration of the Ramazán and its Bairám, the Lent and Easter of the Mohammedans, can wonder that the common people should have learnt to consider that Bairám as the more important of the two.

day. This festival is called by the Turks Kurbán-bairám, the feast of sacrifice, or Khadgilér-bairám, the feast of the khadgís, or pilgrims. The origin of both appellations is explained by the fact that, on the day of its commencement, those holy men slay victims, as the last religious act of their pilgrimage to Mecca. We had a beautiful cloudless sky: our course lay near the shore: we passed, soon after leaving Khaniá, the little flat and barren island where the Venetians had their lazaretto, and, about six miles further on, the village of Haghía Marína, on our left. We met several droves of mules and asses, laden with oil for Khaniá: and, on passing Haghía Marína, saw the village of Plataniá, on a rocky elevation, about half a mile from the shore and a mile before us. Soon after passing it, we crossed its rapid stream, which rises in the White Mountains, and, after flowing between the Rhizite villages of Thériso and Láki or Lákus, runs through a valley formed by low hills, and filled, especially near the stream, with lofty platanes; from which both the village and river obtain their names. Vines twine around most of these platanes, and are of a size unknown in France or Italy, the thickness of many of their stems being that of an ordinary man's waist. These vines are never pruned, and, in consequence of the shadiness of their situation, their fruit does not ripen till after the common vintage: they thus supply the bazár of Khaniá with grapes for the whole month of November, and, I believe, even till near Christmas. The varied scenery produced by these noble plane-trees, in the valley of Plataniá, is very beautiful, and is one of the objects best worth viewing by those who visit Khaniá, and can stay only a short time in the island. The river of Plataniá falls into the sea, nearly opposite the islet of Hághios Theódhoros, where there is good anchorage; and to which merchant vessels, drawing too much water to enter the port of Khaniá, sometimes repair, although the port of Súdha is ordinarily preferred by them. The islet is entirely uninhabited: there was on it, in the time of the Venetians, a small castle, which was heroically defended by its commander Giuliani against the Turks. When summoned to surrender, he replied, that "he had promised to defend the fortress till his last breath;" and, as soon as the Turks carried the place by assault, Giuliani, unable to make any further resistance, fired a mine, the explosion of which destroyed, along with himself and his own small band, all the Turks who had made good their entrance into the fort[2].

As to the river Plataniá, there can be no reasonable doubt about its being the Iardanos of the Odyssey, near the banks of which the Cydonians dwelt[3]: and the elevation on which the village of Plataniá itself is built, may, not improbably, have been the site of the ancient Pergamos or Pergamum, which has been already mentioned incidentally, and the position of which it is not difficult to determine, with tolerable accuracy, from books alone. In the first place it appears, from Servius[4], that Pergamum was near Cydonia; and Pliny[5] places it between Cydonia and the city of Kisamos. If Pergamum was situated here, it is clear that the whole plain running along the shore from this place to the Dictynnaean chain, would form part of the district belonging to it, and thus the Dictynnaean temple of Artemis, which was undoubtedly situated on the modern cape Spádha, might well be spoken of, as it is by Scylax[6], as being to the north of the Pergamian territory.

Various accounts are given of the foundation of the city Pergamum, or Pergamia as it seems to be called by Plutarch. In Virgil's story Æneas is represented

[2] See Emmanuel Mormorio, in the MS. Classe vi. Cod. ci. of Saint Mark's Library, Libro i. fol. 12.

[3] Odyssey, iii. 292.

Ἧχι Κύδωνες ἔναιον Ἰαρδάνου ἀμφὶ ῥέεθρα.

[4] Servius, on Virg. Aen. iii. 133. [5] Pliny, iv. 20.

[6] Scylax, in Geograph. Graec. Min. Tom. i. p. 18. ed. Huds. p. 265. ed. Gail. *Δικτυνναῖον Ἀρτέμιδος ἱερὸν πρὸς βορέαν ἄνεμον τῆς χώρας Περγαμίας.*

as its founder[7]; while Servius, in his commentary on the passage, mentions another account[8], according to which it was built by Agamemnon's Trojan prisoners. Diodorus Siculus, in one of the fragments published from the Vatican Palimpsest, mentions a curse of Agamemnon on the warriors who remained in Crete, and likewise an old proverb of the island, in which the Pergamians are spoken of as the authors of the calamity[9]. I suppose the warriors in question to be those who deserted their Chieftain here, and that the word Pergamians denotes the founders of this Cretan city, rather than the inhabitants of Troy. Velleius Paterculus[10] says that Agamemnon himself, when driven into Crete by a storm, founded Pergamum as well as Tegea and Mycenae. Of the exact situation of these other towns no ancient authors afford any indications. Tegea alone is known by its coins, and the commonest type which they bear is the same as one of those of Cydonia[11]. If the legend of the foundation of these cities by Agamemnon sprang out of the Dorian colonies of Althaemenes, there will be no doubt that those colonies were planted in the western parts of the island[12].

[7] VIRGIL, Aen. III. 133.

> Et tandem antiquis Curetum allabimur oris.
> Ergo avidus muros optatae molior urbis,
> Pergameamque voco.

See also v. 190. of the same book.

[8] SERVIUS, l. c. "Alii dicunt, Pergamum in Creta conditam a Trojanis captivis: qui ex classe Agamemnonis illo erant delati." See HEYNE's "Excursus de Aeneae erroribus."

[9] DIODORUS SICULUS, XXXIII. in Scriptorum Veterum Collect. Nov. Vol. II. ed. Mai. or in Dindorf. Vol. III. p. 109. Ἔθετο ἀρὰν κατὰ τῶν ἀπολειφθέντων εἰς Κρήτην στρατιωτῶν—(cf. Virg. Aen. III. 190. "Hanc quoque deserimus sedem, paucisque relictis Vela damus.) παλαιὰ διαμένει παρὰ τοῖς Κρησὶ παροιμία δι' ἑνὸς στίχου μηνύουσα τὴν νῦν γενηθεῖσαν περιπέτειαν, Αἰαῖ, Περγάμιοι παρά τοι κακὸν ἡγήσαντο.

[10] VELLEIUS PATERCULUS: Lib. I. princip. "Agamemnon, tempestate in Cretam insulam rejectus, tres ibi urbes statuit: duas a patriae nomine, unam a victoriae memoria: Mycenas, Tegeam, Pergamum."

[11] See PELLERIN, Recueil etc. Tom. III. p. 73. Supplem. Tom. I. p. 52. ECKHEL, Num. Vet. p. 141. SESTINI, and MIONNET, Suppl. IV. p. 343.

[12] CONON, 47. EUSTATHIUS, on IL. II. p. 313. HOECK's Kreta, Vol. II. p. 429. and THIRLWALL's History of Greece, Vol. I. p. 279.

As the Cnossians long boasted of the tombs of two celebrated ancient heroes, Idomeneus and Meriones[13], so the Pergamians used to shew that of the great Spartan lawgiver Lycurgus, who, according to accounts in Plutarch's life of him, died in Crete, and was buried here[14].

Leaving the district of Plataniá, I soon arrived at the village of Ierámi, in and about which I lost a good deal of time in making fruitless inquiries for existing remains of antiquity. I think my search was sufficiently diligent to warrant me in asserting, that there are positively no ancient remains whatever in the immediate neighbourhood of the place. I next rode to a hamlet called Pýrgos, a little further on, where my researches were equally unsuccessful. Unfortunately I had extracted a passage from the work of a fellow-countryman[15], wherein it is said, after speaking of Cydonia, "The ruins of this ancient city are to be seen on the site of Ierámi." The very difficulty, which the existence of any considerable vestiges of an ancient city at Ierámi would have caused in the way of what I have endeavoured to establish, respecting the probable sites of Aptera, Cydonia and Polyrrhenia, made me so anxious to verify this assertion, that I passed Plataniá without making any search for remains of an ancient city there.

Neither ancient authors, nor modern travellers, as far as I am aware, afford any ground or colour for Dr Cramer's assertion, and I know not how to account for it, unless he has taken for granted the existence of these imaginary ruins on the authority of Lapie's map[16].

13 Vol. I. p. 206.

14 PLUTARCH, Lycurg. c. 31. (Vol. I. p. 234. ed. Reisk.) *Τίμαιος δὲ καὶ Ἀριστόξενος ἐν Κρήτῃ (λέγουσι) καταβιώσαντα, καὶ τάφον Ἀριστόξενος αὐτοῦ δείκνυσθαί φησιν ὑπὸ Κρητῶν τῆς Περγαμίας περὶ τὴν ξενικὴν ὁδόν.*

15 Dr CRAMER's Geographical and Historical Description of Ancient Greece, Vol. III. p. 366.

16 As seems to have been done by Mr GAIL, Geograph. Graec. Min. Vol. II. p. 581. "Reperiuntur hod. rudera Cydoniae in litore ipso. Olim perperam veterem Cydoniam in hodierna Canea reponebant; sed jacent rudera illa ulterius ad occidentem fluvii hod. *Platania*."

With whomsoever the mistake may rest, this is not simply one of those numerous instances in which Lapie has merely given wrong names to ancient sites, for here the error will be equally great to whatever ancient city the remains are supposed to have belonged; and we may therefore refer this to that class of sites where, as in the case of Arcadia, ruins are supposed to exist in given spots, and then the name of some unlucky ancient city is attached to them, and thus the city acquires, in the present day, a new local habitation, and the imaginary ruins obtain a name.

Half an hour's riding from Pýrgos brings us to a river which we cross: the monastery of Goniá is about three miles before us at some little elevation above the shore; and we have for some time seen "The convent's white walls glisten fair on high." The olive-trees on our left, for a considerable distance, are entirely uncultivated. We cross another small river, just before winding round the south-west corner of the gulf, and, in a few minutes more, arrive at the monastery, situated on the side of rocky hills, and only about a hundred paces from the sea. The Hegúmenos is absent: the senior Patéras however does the honours, shewing us into a very comfortable room, and causing dinner to be prepared. The wine which is produced fully supports the credit of monastic institutions.

Not only in Crete, but elsewhere, the wine of the Greek monasteries is mostly excellent, and this too in countries where its quality is generally bad. At the same time, the sobriety and abstemiousness of eastern manners have preserved the Greek church from the scandal of having the sounds of revelry and debauchery heard within the precincts of its monasteries, as used to be the case in Catholic countries.

> "There Venus sat disguised like a Nun,
> While Bacchus, clothed in semblance of a Friar,
> Poured out his choicest beverage, high and higher,
> Sparkling."

H. Schranz del. L. Haghe Lith.

GONIA and the WHITE MOUNTAINS

Such scandal as the Roman Catholic church has endured, from the time of Boccaccio downwards, on account of the profligacy of her monks, the Oriental church has certainly escaped: and yet, even here, the vices inherent in the very principles of an institution which opposes the first great command of God and nature, by imposing celibacy on all its members, have been sufficiently developed to have called for the ineffectual prohibitions of a General Council in the early ages of Christianity[17], to have become the theme of popular songs at the present day[18], and, at last, to have induced the Bavarian government of Greece, actually to put down the monastic order in all parts of the new kingdom, and to apply the revenues, which these useless monks consumed, to the purposes of education.

Notwithstanding the Hegúmenos's absence we fared well: and, without our having transgressed the limits of reason and moderation, he found, on his return in the evening, that our time after dinner had not been unemployed. The most serious subject of our conversation was the affair of Murniés, and I soon learnt, that every absurd belief which was entertained by the body of the peasants there assembled, is also shared by my reverend friends: for instance, it is quite impossible to disabuse them of the idea that Mehmét-Alí's sole object in establishing schools at Neokhório, was to seize on all the children some fine day, to put them on board a ship of war lying in the bay of Súdha, which was conveniently near, and to carry them off to Egypt. Their ideas respecting the manner in which the Pashá made his terms with the English Admiral and the Consuls, are still more absurd. In these unhappy countries, the Superior of the most liberally endowed mo-

[17] As is observed by GIBBON, c. XXXVII. "The seventh general council prohibits the erection of double or promiscuous monasteries of both sexes; but it appears from Balsamon, that the prohibition was ineffectual." Several of these promiscuous convents still exist in Crete: e. g. Haghía Hodegetría, and Pezanés.

[18] See above, Vol. I. p. 146.

nastery is barely removed, in point of education and of knowledge, one degree from the condition of the poorest peasant.

April 22.

The church of this monastery contains a greater number of those paintings with which the Greeks love to adorn the interior of their places of religious worship, than is usually met with in Crete. They were all sent to Trieste at the outbreaking of the revolution, and thus escaped destruction. One of them has the date 1642. Two of the most remarkable are, the history of Joseph, which contains a great number of groups, exhibiting all the chief events of his life, and another wherein the Virgin and child are represented in a kind of tub, out of which are flowing two streams[19]. The Mohammedans did not burn this church, and even left uninjured all the carved wood with which it is adorned; an instance of moderation rare on their part, and which most assuredly was never shewn by the Christians in any mosque in the island. After visiting the church I enter the refectory: on either side of its entrance is a Corinthian column: there is in fact here, as in most of the Venetian edifices which remain in Crete, some pretension to architectural design and ornament. The interior of the refectory is about twelve paces long and five or six wide; and has a table with benches on one side of it: it is large enough to remind me, by its size as well as by its furniture, of a small college-hall at Cambridge.

In its possessions this monastery has of course suffered, like all the others, by the war. It contained

[19] This painting is called Η ΖΩΟΤΟΚΟΣ ΠΗΓΗ. Whether these were streams of milk or water, I will not now take upon myself to decide, since I do not recollect how they were supplied. If they were milk, my reader will perhaps remember that the copiousness of this source in a Christian church nearer to us than that of the Greeks, is alluded to by SWIFT, when he makes lord Peter swear "that he had a cow at home, which gave as much milk as would fill three thousand churches, and, what was yet more extraordinary, would never turn sour." (Tale of a Tub, §. IV.)

formerly twenty patéras, fifteen kalógheri, four or five deacons[20], and between fifty and sixty men-servants[21]. Its eight metókhis were then cultivated by its men-servants, under the superintendance of a patéras residing on each farm. At present the monastery is unable to till its own lands, and therefore lets them to a cultivator, receiving only one half of the produce, and supplying the farmer with oxen: thus the half of the produce received by the monks is generally a very small one.

Before quitting our kind and hospitable hosts we partake of a very palatable breakfast, chiefly consisting of a kind of piláv, made of artichokes and rice, at the same time doing justice, notwithstanding the early hour, to the excellent wine which they again produce. I find one of the priests a perfect oracle on the subject of the population of every village of the eparkhía of Kísamos. On making inquiries for the remains of Dictynnaeon, I was told that "about three miles from the extremity of Cape Spádha, on its eastern side, are remains of an ancient Greek city at a place called Kantsilliéres, which is not surrounded by walls, but had many houses in former times." I did not think it worth while to spend a whole day in going to see the slight remains which exist of this city, since its position is clearly ascertained. There is no doubt that this Dictynnaeon is the Dictynna of Pomponius Mela[22] and the Dictamnon of Ptolemy[23].

The following is Pococke's account[24] of the remains found there. "The ancient remains of this place are chiefly on a small height over the west end of the bay, and on each side of two rivulets, which meet just before they fall into the sea; most of them are roughly built

[20] *Διάκονοι*, commonly corrupted into *διάκοι*.

[21] Men-servants, *δουλευτάδες*, the plural from *δουλευτής*.

[22] POMPONIUS MELA, II. 7. 12. and TZSCHUCKE's Notae Criticae, in Tom. II. Part. II. p. 551.

[23] PTOLEMY, Geograph. p. 91.

[24] POCOCKE's Description of the East and some other countries, Vol. II. Part I. pp. 244–245.

of the grey marble of the mountains which are on each side: one building resembles a church, and has some ancient brick-work about it. On a height to the south of the bay, there are some pieces of grey marble columns, and four oblong square cisterns sunk into the ground and contiguous, as if they had been under some great building. I observed that in the middle they were sunk lower like square wells, and lined with brick, with a design, I suppose, to receive a greater quantity of water; and below these, on the side of the hill towards the town, there are remains in some of the walls of earthen pipes, by which one may suppose the water was conveyed down from the cistern, the torrents below being dry in summer. Among these ruins, which were probably an ancient temple, I saw a fine pedestal of grey marble three feet square; it had a festoon on each side, and against the middle of each festoon a relief of Pan standing; the whole was finely executed; it is probable that this was either an altar or the pedestal of a statue erected to that deity in this temple, which probably was dedicated to the nymph Dictynna; Strabo mentions the Dictynnaean temple in this place. Some years ago they found a statue here of white alabaster, but having a notion that such pieces of antiquity contain gold in them, the fishermen broke it in pieces; I brought away a foot of it, which shews very distinctly all the parts of the ancient sandal."

KISAMO-KASTELI.

CHAPTER XXIII.

PLANE-TREES. VIEW OF MOUNT IDA. PLAIN OF KISAMOS. SUPPOSED CHANGE IN THE SUN'S USUAL PLACE OF RISING. ASPECT OF ANCIENT TEMPLES AND OF CHRISTIAN CHURCHES. INFLUENCE OF THE ORIENTAL SUN-WORSHIP ON THE MODE OF PRAYING, AND ON THE ARCHITECTURE OF THE EARLY CHRISTIANS. KEBLAHS OF THE JEWS AND MOHAMMEDANS. VESTIGES OF METHYMNA. PROBABLE SITE OF RHOKKA. REMAINS OF KISAMOS. VISIT TO THE ALBANIAN COMMANDANT OF KISAMO-KASTELI. ARRIVAL AT UPPER-PALAEOKASTRON. CRETAN WILD-ASPARAGUS.

April 22.

We start at a quarter before nine o'clock, and, after passing Agribilianá, and traversing groves of olive-trees, which are almost entirely uncultivated, we leave the village of Spílaea on our left, and arrive at a fountain shaded by two plane-trees, one of them of most majestic dimensions. For some little distance further we continue to ascend, and, until we cease to do so, see behind us, not only the Akrotéri but the whole of the snow-capt eminence of Mount Ida, although at a distance of about sixty miles. After crossing this ridge, we pass the village of Nokiá, which is a little on our right, and soon come in sight of the gulf of Kísamos. In about an hour from Nokiá we pass a fountain, and, a few minutes afterwards, see the plain of Kísamo-Kastéli, which is about four miles long and a mile and a half broad. It is chiefly covered with standing corn: there are, however, also considerable patches of olive-trees. The Panopolitan poet has thought this plain worthy of being expressly mentioned[1]. The Kastéli is situated near its further extremity. After advancing a mile more, we are just above Nopía, which is on the extreme eastern edge of the plain. A river running to the west of this village separates it from the church of Hághios Gheórghios. We were now accompanied by a Greek called Antónios Kharadhákes of Nopía, who had been overtaken by us half an hour before, and had immediately volunteered his services to conduct me to certain remains of antiquity close to this church, which seems to have been built on the site of an ancient temple. Part of the foundations of the older edifice may still be traced. In the large masses of brick and mortar, which are lying about near the modern building, there are seen three earthen pipes, such as one ordinarily finds only in cisterns.

This Greek chapel stands nearly due north and south, instead of east and west as is usual in all orthodox Greek as well as Roman Catholic and Protestant churches. This architectural peculiarity of the building puzzled my Nopían guide quite as much as myself. He asked me whether in those times, for he transferred the antiquity of the ancient edifice to the comparatively modern erection, the sun used to rise in the south: it must have been so, he thought, because the Christian, as he stands before the altar in a church, always looks towards the rising sun.

[1] NONNUS, XIII. 237.

Καὶ δάπεδον Κισάμοιο.

The Greek's idea calls to mind Herodotus's mention of what he heard in Egypt, about the change which had happened, in ancient times, in the sun's usual places of rising and setting[2].

Ne is that same great glorious lamp of light,
That doth enlumine all the lesser fyres,
In better case, ne keeps his course more right,
But is miscaried with the other spheres.

And if to those Egyptian wizards old,
Which in star-read were wont have best insight,
Faith may be given, it is by them told
That since the time they first tooke the Sunnes hight,
Foure times his place he shifted hath in sight,
And twice hath risen where he now doth west,
And wested twice where he ought rise aright[3].

I have never before noticed an instance, in any part of Greece or Turkey, of a church not standing due east and west[4]. In England there is sometimes a deviation, determined by the place where the sun rises on the day of the saint to whom the church is dedicated. This peculiarity in our old ecclesiastical architecture has been celebrated by a living poet[5].

A passage of Ezekiel shews how greatly the Jews abominated the oriental custom of turning the face to the east in prayer[6]. The Jews themselves seem to have turned towards the sanctuary, even in the time of David[7];

[2] HERODOTUS, II. 142. Ἐν τοίνυν τούτῳ τῷ χρόνῳ τετράκις ἔλεγον ἐξ ἠθέων τὸν ἥλιον ἀνατεῖλαι. See also BUTLER, Hudibras, Canto III. v. 865. foll.

[3] SPENSER, Faerie Queene, Book V. Introd. Stan. 8.

[4] No doubt such may be found. Colonel LEAKE, Travels in the Moréa, I. p. 297. speaks of the church of Asómatos, near Cape Matapán, which "instead of facing to the East, as Greek churches usually do, faces south-eastward, towards the head of the port, which is likely to have been the aspect of the temple," (of the Taenarian Poseidon.)

[5] WORDSWORTH.

[6] EZEKIEL, VIII. 16—18. The ruins of a temple called Ghebri Bena, the temple of the Ghebers, were visited by MR KEPPEL, Journey from India to England, Vol. I. Ch. X. He observes, "The gradual slope of the plain to the west would indicate that on that side was the ascent to the temple."

[7] PSALMS, V. 7. XXVIII. 2.

and, when Solomon's temple was built, it became their keblah[8]. Hence, during the captivity, they used to turn towards Jerusalem, then to the west of them, in their prayers. Thus Daniel, when at the court of Darius, "went into his house, and his windows being open in his chamber, *toward Jerusalem,* he kneeled upon his knees three times a day, and prayed and gave thanks before his God as he did aforetime[9]."

This Jewish observance does not appear to have been adopted by the early Christians, among whom the Persian custom of turning towards the rising sun every where prevailed[10]. Directions are given to build all churches so as to stand towards the east in the Apostolical Constitutions[11]. Origen[12], Clemens[13], Basil[14], and other ecclesiastical writers, attempt to assign *reasons* for the practice: some of them, as Tertullian, consider the sun the figure of Christ: others say that, since Paradise was situated in the east, we mean our looks in that direction as a prayer that we may be restored to the place from which Adam was expelled: others, as if they meant to confine Christianity to the *west* of Jerusalem, tell us that we look to the scene of Christ's abode and

[8] 1 KINGS, c. VIII. See especially v. 48. 2 CHRONICLES, VI. 24. 38. and JONAH, II. 4. 7. The entrance of the temple was on its eastern side, in order that those who went into it to pray might have their *backs* towards the rising sun. ANASTASIUS, Quaest. XVIII. in Script. (quoted by Potter, on Clem. Alex. p. 857.) Ἐτέτραπτο δὲ πρὸς ἕω, ἵνα οἱ προσευχόμενοι μὴ τὸν ἥλιον ἀνίσχοντα προσκυνῶσιν, ἀλλὰ τὸν ἡλίου δεσπότην.

[9] DANIEL, VI. 10.

[10] ORIGEN, Opera, Vol. II. p. 284. ed. Par. 1733. (Hom. V. on Numbers.) "Quod ex omnibus coeli plagis ad solam orientis partem conversi orationem fundimus, non facile cuiquam puto ratione compertum." TERTULLIAN says: "Ad Persas si deputabimur, licet solem in linteo depictum non adoremus —inde suspicio, quod innotuerit nos ad Orientis regionem precari." Bishop KAYE says, on Tertullian, p. 124. (compare p. 408.) in speaking of the enemies of the Gospel, "We cannot be surprized at their believing that the Sun and the Cross were objects of worship in the New Religion."

[11] CONSTITUTIONES APOSTOLICAE, II. c. 57. Καὶ πρῶτον μὲν ὁ οἶκος ἔστω ἐπιμήκης κατ' ἀνατολὰς τετραμμένος.

[12] ORIGEN, de Orat. Tom. I. p. 270.

[13] CLEMENS ALEXANDRINUS, Strom. VII. p. 856.

[14] BASIL, de Spiritu Sancto, c. XXVII. p. 56. a. ed. Paris, 1730.

sufferings on earth: and this last notion, that we thus turn because

> Mindful of him, who in the orient born,
> There lived, and on the Cross his life resigned,

destitute of all truth as it undoubtedly is, if the origin of the observance is considered, has become very generally prevalent. Christian authors are likewise found who venture to assert that the custom was an institution of the apostles[15]. It appears from Philo-Judaeus, that the Therapeutae prayed in the same manner: and the observance being so strictly Christian, it forms one of the arguments by which it has been attempted to prove that the Therapeutae were Christians[16]. The early practice still prevails, more or less, in all countries inhabited by Christians to the east of the Holy Land, as well as in those to the west of it.

The Greeks and Romans seem also to have turned to the east in offering prayer and sacrifice, and thus the oriental observance was easily continued among the Christian converts from Paganism throughout the Roman empire[17].

[15] AUCTOR Quaestion. et Respons. ad Orthodoxos, Resp. 118. Τὸ δὲ ἔθος, παρ' ὧν εἴληφεν ἡ ἐκκλησία τὸ εὔχεσθαι, παρὰ τούτων εἰληφέναι καὶ τὸ ποῦ εὔχεσθαι, τοῦτο δὲ παρὰ τῶν ἁγίων ἀποστόλων. ATHANASIUS, Quaest. ad Antiochum, 16. (quoted by Potter, l. c.) Οἱ μακάριοι ἀπόστολοι κατὰ ἀνατολὰς τὰς τῶν Χριστιανῶν ἐκκλησίας προσέχειν πεποιήκασιν.

[16] Le Livre de Philon de la vie contemplative traduit de l'original Grec, avec des observations, où l'on fait voir que les Therapeutes dont il parle étoient Chrétiens. Paris, 1709. pp. 200—202. The chief establishment of these Therapeutae was in lower Egypt: an account of them is given in NEANDER, Geschichte der C. R. u. K. Vol. I. Part I. p. 78. foll. They were undoubtedly a sect of Jews, and not Christians.

[17] VITRUVIUS, IV. 5. "*Signum* quod erit in cella collocatum spectet ad *vespertinam* coeli regionem." See also IV. 9. and VIRGIL, Aen. VIII. 68.

> Surgit, et aetherii spectans orientia Solis
> Lumina, rite cavis undam de flumine palmis
> Sustulit, ac tales effundit ad aethera voces.

And Aen. XII. 172.

> Ad surgentem conversi lumina solem,
> Dant fruges manibus salsas, et tempora ferro
> Summa notant pecudum, paterisque altaria libant.

LUCIAN, Necyomant, §. 7. Tom. I. p. 465. ed. Hemst. Other passages are indicated by LOMEIER, de Vet. Lustrat. p. 316. fol. ed. 1681.

Nevertheless it is undoubted that this ancient custom was greatly changed, although the alteration introduced was equally consistent with the principles of the sun-worship in question. Hyginus mentions the revolution introduced into sacred architecture by building temples, the *entrance* into which was at their *eastern* extremity[18]; and it appears, from Lucian[19] and other authors, that those eastern doors were thrown open, so as to admit within the sacred edifice the first beams of the rising sun.

The custom of thus constructing temples with an eastern aspect became very general; and, in later times, the form was sometimes transferred from the heathen to the Christian edifice. The temple of Apollo at Delphi[20] and the cathedral of St Peter at Rome were thus equally turned towards the rising sun.

It is quite certain that the observance of the great celestial luminary, adopted by the Christians, has at times degenerated into something little better than the Persian worship, of which Herodotus and other ancient authors speak. I myself once met with an ignorant Greek, who told me that the great difference between Christianity and Mohammedanism consisted in this, that the Christian worships the sun, and the Mohammedan the moon[21]: and we learn from the mouth of one of

[18] HYGINUS, de agror. limit. constit. p. 153. in the Rei Agrariae Auctores, ed. Amst. 1674. "Antiqui architecti in occidentem templa spectare recte scripserunt: postea placuit omnem religionem eo convertere, ex qua parte coeli terra illuminatur."

[19] LUCIAN, de domo, Tom. III. p. 193. *Τὸν ἥλιον ὑπερκύψαντα εὐθὺς ὑποδέχεσθαι καὶ τοῦ φωτὸς ἐμπίπλασθαι ἐς κόρον, ἀναπεπταμένων τῶν θυρῶν, καθ' ὃ καὶ τὰ ἱερὰ βλέποντα ἐποίουν οἱ παλαιοί.* See also PORPHYRY and DIONYSIUS THRAX, quoted by LAKEMACHER, Antiq. Graec. Sacr. Part. I. c. VI. §. 4.

[20] LEAKE, Travels in Northern Greece, Vol. II. p. 562.

[21] *Ἐγὼ προσκυνῶ τὸν ἥλιον, καὶ ὁ Τοῦρκος τὸ φεγγάρι.* His explanation of this strange view of the matter was that all churches have the altar at the east end, and that every Christian turns to the rising sun, &c., while "the Turk reckons by the moon," and on some of his festivals celebrates her rising by the discharge of fire-arms, and exhibits the crescent on his standard: *καὶ ἔτσι ὁ Τοῦρκος προσκυνεῖ τὸ φεγγάρι.* RELAND, de Rel. Mahom. p. 163. says: "Confirmat hanc opinionem, Lunae imaginem esse insigne Mohammedanorum, quasi hoc ex antiquo Planetarum cultu apud ipsos remanserit."

the successors of St Peter, that this adoration of the bright orb of day was practised by many Christians of his time[22]. They even turned their backs on the altar[23], in the most splendid temple which Christian piety has ever erected, that they might bow down in their Pagan adoration of the rising sun.

Plutarch and Aelian both mention an ancient opinion, from which alone it might be inferred how essential a part of the ancient religious worship it was thus to turn to the east. The act of the elephant, in stretching out his proboscis towards the rising sun, was regarded as one of religious adoration of the Deity[24].

The first direction given to the Mohammedans by the founder of their religion was worthy even of a true prophet: "To God belongeth the east and the west; therefore whithersoever ye turn yourselves to pray, there is the face of God[25]." After this Mohammed bade his followers turn towards Jerusalem; but, at last, adopting the principle, while he abandoned the practice, of the Jews, he enjoined on them to turn in the direction of the Caaba at Mecca[26]: "Follow ye therefore the religion

[22] VOSSIUS, de origin. et prog. Idolol. II. 3. Quo pertinent haec Leonis Papae, Serm. VII. de Nativ. "De talibus institutis etiam illa generatur impietas, ut sol, inchoatione diurnae lucis exsurgens, a quibusdam insipientioribus de locis eminentioribus adoretur: quod nonnulli etiam *Christiani* adeo se religiose facere putant, ut—*converso corpore ad nascentem se solem reflectant, et, curvatis cervicibus, in honorem se splendidi orbis inclinent.*"

[23] ST LEO, l. c. "Priusquam ad beati Petri apostoli basilicam, qui uni Deo vivo et vero est dedicata, perveniant; superatis gradibus, (it must be remembered that, as is mentioned in the text, the entrance of St Peter's is at its eastern extremity, towards the eternal city,) quibus ad suggestum arae superioris adscenditur, converso corpore, ad nascentem se solem," &c.

[24] PLUTARCH, Tom. II. p. 972. b. Ἱστορεῖ δὲ καὶ εὐχῇ χρῆσθαι θεῶν τοὺς ἐλέφαντας ἀδιδάκτως, ἁγνιζομένους τε τῇ θαλάσσῃ, καὶ τὸν ἥλιον ἐκφανέντα προσκυνοῦντας, ὥσπερ χειρὸς ἀνατάσει τῆς προβοσκίδος· ὅθεν καὶ θεοφιλέστατόν ἐστι τὸ θηρίον. AELIAN, N. A. VII. 44. Τὸν ἥλιον ἀνίσχοντα προσκυνοῦσιν ἐλέφαντες, τὰς προβοσκίδας εὐθὺ τῆς ἀκτῖνος ὡς χεῖρας ἀνατείνοντες, ἔνθεν τοι καὶ τῷ θεῷ φιλοῦνται.

[25] An Arabic poet elegantly speaks of the Deity as his Keblah: see RELAND, de relig. Mahom. Lib. II. §. X. p. 175.

[26] As the Arabs did before his time: RELAND, de rel. Mahom. p. 271.

of Abraham, for he was no idolater. Verily the first house appointed unto men to worship in, was that which is at Mecca, a direction to all creatures." Again, "We have seen thee turn about thy face towards heaven with uncertainty, but we will cause thee to turn thyself towards a Keblah that will please thee. Turn therefore thy face towards the holy temple of Mecca; and wherever ye be, turn your faces towards that place." Mosques are not built so as to point out the direction to Mecca, as a church ordinarily points out the east, but it is determined within the mosque by a niche which they call Al Mehrâb; and without, by the situation of doors opening into the galleries of the steeples[27].

The Mahommedans also use, especially in India, small compasses in which the magnetic needle is concealed under the wings of a bird, so placed, that the bird's flight may always appear to be towards Mecca. One of these compasses has been shewn to me by the Master of Pembroke College, whose brother, Mr Ainslie, brought it from Benares. On it the bird's flight is to the west-south-west, a direction which is sufficiently explained by the fact that the instrument was meant for India. Mr Ainslie tells me that, in India, magnetic needles, serving in a similar way to point to the Holy City, and set in rings, are very common among the Mohammedans. Thus, even where he has no mosque in sight, the devout Musulman is enabled to turn toward Mecca, in those daily performances of his religious exercises, the punctual observance of which, at all times and under all circumstances, no traveller in the east can fail to notice.

I may add, that at the south-east corner of the chapel of Hághios Gheórghios is a fragment of circular walling. I learn from Antónios that the Greeks dug here, during the revolution, "and found a woman and

[27] See SALE.

a child of marble, which they sent to Anápli:" there were twenty-five of them who excavated, and they afterwards received nine dollars a-piece, the amount of each man's share of the sum obtained for the statues. A Melian of the name of Ioánnes, called, from his profession of dealer in antiquities, Antíka-Yánnis, also came and dug here afterwards. I am told marvellous stories about the supernatural appearance of a negro, who, when the excavators had discovered a certain entrance, was seen to stand with a drawn sword, forbidding them to proceed with their work, and who, in fact, deterred them from doing so.

Before I crossed the streamlet to visit the remains at this church, I found the Albanian commandant of Kísamo-Kastéli seated smoking his pipe under a tree just by the village, and waiting till some of the villagers prepared his dinner: it was now nearly noon. He wondered what could be my object in asking questions about the country, as he found out, both from Maniás and my Nopían guide, that I did. I succeeded with a little difficulty in making him understand the reasonableness of my endeavour to note down something of places far distant from my home, in which I had spent several months, and which, in all probability, I should never see again.

Having finished my survey of the remains, at the church of Hághios Gheórghios, I went to a small hill about a quarter of a mile to the south-south-east of the church, and south of the village of Nopía. On the north-west side of this hill are two projecting square towers, each about eight paces in width and with a curtain of thirty-three paces in length between them. Round the rest of the hill, which is only a few hundred paces in circumference, are remains of a wall six feet thick. It may be observed, that this hill is entirely commanded by the loftier one about two hundred paces off, on the side of which is situated the village of Nopía. The stones of the wall and towers are of various sizes.

The fort, for such it must have been, may, as far as the style of building is concerned, have been a work either of the Greeks or Saracens in the ninth century, of a later period of the Byzantine empire, or of the Genoese at the beginning of the thirteenth[28].

The remains existing at the chapel of Hághios Gheórghios, though slight, are yet sufficient to indicate the existence of an ancient city on the spot. Now we shall see by and by that there was a city Rhokka, where Artemis Rhokkaea was worshipped, a little inland, to the south of this place; and a curious story told by Aelian[29], respecting a remedy for hydrophobia[30] discovered by a Cretan fisherman, shews plainly that there was a village called Methymna on the shore, and

[28] POCOCKE's account of these places is singularly indistinct: he says, "Over this river, on an advanced rock, there are ruins of a house and chapel called Nopeia; about them are the remains of a strong-built wall, five feet thick, as if it had been part of a fortified castle," (p. 245.) I suppose him not to have seen the ruins at the chapel of Hághios Gheórghios, although he contrives to speak of all three, of the village of Nopía, of a chapel, and of these walls.

[29] AELIAN, N. A. XIV. 20. His account is that the old fisherman had caught many hippocampuses, and his sons were bitten by a mad dog. Οἱ μὲν οὖν, he proceeds, ἔκειντο Μηθύμνης τῆς Κρητικῆς πρὸς ταῖς ᾐόσιν—ἔστι δὲ αὕτη κώμη, ὥς φασιν—οἱ δὲ θεώμενοι συνήλγουν τῷ πάθει—οἱ δὲ ἐς τῆς Ῥοκκαίας οὕτω καλουμένης Ἀρτέμιδος ἄγειν, (προσέταττον,) καὶ αἰτεῖν ἴασιν παρὰ τῆς θεοῦ.

[30] Mad dogs used to be common in Crete in ancient time: CAELIUS AURELIANUS, de morb. acut. III. 15. "Haec insula (Creta) aliorum venenosorum animalium difficilis, atque pene libera, sola canum rabie vexatur frequentissime." Several instances of canine madness occurred during my recent residence at Khaniá; and the Council, (Τὸ Συμβούλιον,) or, in other words, its President Raffét-efféndi, has at length given an order, in compliance with the requests of the European Consuls, that all dogs found running about the streets be destroyed. The prejudices of some of the Mohammedans of the old school were very difficult to overcome. The peculiar tenderness of Mohammedanism, towards all the inferior tribes of animals, is too well known to require any observation. Nothing can be more unfounded than SONNINI's assertion, Tom. I. p. 429. "La race (des chiens de Candie) est abâtardie, sur-tout depuis que les Turcs, *grands ennemis des chiens*, se sont emparés de cette belle contrée." On the Cretan breed of dogs, see Vol. I. p. 33. What I have said there, on the dogs of Molossis and Chaonia, throws light on a passage of ARISTOPHANES, Thesm. 416.

Καὶ προσέτι Μολοττικοὺς
τρέφουσι, μορμολύκεια τοῖς μοιχοῖς, κύνας.

near the temple of Artemis Rhokkaea. I think there is no doubt of this being the site of Methymna.

As to Rhokka, I did not visit the spot where I now suppose it to have been situated; for not having had an opportunity of consulting Pococke's Travels, after I had determined to spend as many months as might be requisite in the examination of this island; and, what was a much greater misfortune, not having succeeded in obtaining a copy of Professor Hoeck's valuable work, this site eluded my researches.

Those alone, who have travelled in Greece or Turkey, can duly appreciate the difficulty of discovering unvisited vestiges of ancient cities. I used to spend, I may safely say, hours daily, during all my travels in Crete, in making diligent inquiries for any traces which might indicate the existence of such remains; and thus I have but seldom failed to discover them, where I had any ground for supposing that they ought to exist in the neighbourhood of places which I visited.

The site of the city Rhokka, which thus escaped me, was seen by Pococke, on his return from the remains of Polyrrhenia to Khaniá. He says[31], "Another inland town was Artacina, which might have been at a place called Rocca, though Ptolemy places it more to the south; it is a small high rocky hill, on the top of which are the remains of some buildings: there are about three or four rooms, which the people say belonged to the ancient Greeks.—To the west of this place there is a river called Tiphlosè.—About a league to the north-east of Rocca is a village called Episcope, where there is a church still entire, and the bishop of Chisamo thinks that it is his cathedral." That Hyrtakina was further to the south than the remains visited by Pococke will be seen hereafter. This site is, I have no doubt, that of Rhokka, where, according to Aelian[32], there was a temple of Rhokkaean Artemis, which of course would

[31] POCOCKE, Tom. II. p. 247. [32] AELIAN, N. A. XII. 22.

be but a short journey distant from the city of which we see vestiges at Nopía. The river here spoken of by Pococke is the Typhlós, which runs about a mile to the westward of Nopía, so that Rhokka is a little to the south of this place. It is plain from Pococke's journal that it cannot be very far inland, for he says, after describing the remains of Polyrrhenia, which he supposed to be those of Aptera, that on finding he could get no other information than he already possessed respecting Phalasarna, Khersonesos and Inakhorion, he went no further that way. He then adds, "Strabo observes, that the island is twenty-five miles broad at the west end, and Ptolemy makes it thirty." It is therefore evident that he did not go to the southward of Polyrrhenia, but returned towards Khaniá, taking an inland road to avoid retracing his steps along the shore. Thus it appears clear and undoubted that the ruins he saw are very nearly where I have placed them on my map.

I am disposed to believe that the remains of Rhokka were also visited by Buondelmonti. He mentions the river Typhlós as being near Kísamos, and then adds, that in going southward, after crossing the Typhlós, I suppose, he saw a new church in which were ancient marble and porphyry columns: most likely remains of Artemis's temple[33].

I should not have thought it worth while to enter into so careful an investigation of Pococke's words, had not Professor Hoeck, whose opinions always deserve the greatest attention, placed Rhokka and Methymna near the western shore, considerably to the southward of Sfinári.

[33] BUONDELMONTI, in the Creta Sacra of CORNELIUS, Vol. I. p. 6. After mentioning the river Typhlós, "Cumque versus Austrum per medium Insulae proceditur, nova ecclesia invenitur, in qua *columnae porphireticae et marmoreae antiquae* existunt." In his more verbose account at p. 88. he speaks of their arrival at the modern church: "dum intravimus versus Pavimentum, respice, dixit, cognosce Musaicum, quod jam per tot saecula a pluviis et terrae humectationibus illaesum est, atque totam hanc vineam cum multis hedificiis porfireisque marmoreis comprehendit in unum." This

modern

We leave Nopía, the site of Methymna, at forty-five minutes past one: pass the village of Dhrapaniá, cross the river Typhlós, and then, passing the village of Kurvalónes, arrive at fifty minutes past two at the river of Kamára, where I observe, on one side of the streamlet, the massive supports of a bridge which no longer exists. The village of Kamâra had formerly forty or fifty houses, it is now a complete heap of ruins: all its inhabitants were Mohammedans, and, like most of their correligionaries of the district, had taken refuge in the Kastéli, where they were blockaded by the revolted Greeks, and where nearly all of them fell victims to the plague. Half a mile further we arrived at Kísamo-Kastéli, having observed, just before entering the town, some ancient arched caves in the Turkish cemetery: they are almost entirely under ground.

In and about the town are fourteen or fifteen fragments of shafts of marble and granite columns, an Ionic capital, another column's base, and some remains of walls. Other vestiges, which sufficiently prove that there once existed here a flourishing and important city, are seen scattered over nearly half a mile round about the Kastéli. Four centuries ago Buondelmonti saw many columns and other remains of a considerable ancient edifice within the modern town[34]. The principal ruins now are at some little distance to the southward of Kísamo-Kastéli, and consist of a large mass of walls fallen in confusion: its outer facing is of brick-work, and its interior consists of small stones and mortar: and what would appear to have been an aqueduct, of which, however, only the

modern coast-describer, proceeding in his periplus, passes along the sandy shore, and arrives "ad flumen Napoliam non magnum," or as we read at p. 89. "ad non magnum Nopiliam flumen," by which I understand the river of Nopia. He describes the mountains of Cape Spádha as abounding in caves, pp. 6. and 89.

[34] BUONDELMONTI, in the Creta Sacra of CORNELIUS, Vol. I. p. 6. "Antiqua Chissamospolis oppidum album videtur, in cujus medio fons uberrimus dulcis aquae, et palatium cum multitudine columnarum jam fere prostratum cernitur."

summit of the arch is visible: the ground about is evidently a good deal higher than the ancient level[35].

I paid a visit to the Albanian commander of the fortress, who had passed as we were examining the ancient remains. He received me with the ordinary civilities. On my presenting him the Pashá's letter, I found out, what did not at all surprise me, that the Arnaut could not read. He, however, sent immediately for the Kadí, who performed the office of secretary, and with whom and another Cretan Turk of this place, I had a long conversation about the blockade of the Kastéli, and the sufferings of the Mohammedans within the fort, where the plague was then raging most fiercely. They were at first near eight hundred men, and were barely seventy when they capitulated. This place consists of the castle, and a small town also walled in: they are not, both together, much more than half a mile in circumference.

On leaving Kísamo-Kastéli we proceed to the south-east: the olive-trees, which we pass on either hand, are more than half of them uncultivated, and the ground below them is covered with furze and brushwood: we arrive at Lower Palaeókastro in about half an hour, and a continued ascent of half an hour more brings us to Upper Palaeókastro. I observed remains of ancient walls before arriving at the village, and, on entering it, noticed a curious tower.

Being still distant many days from the hour of our deliverance from the Greek Lent, I was astonished to see the peasant who was our host for the night produce some meat, and make preparations for cooking it. On

[35] I did not hear of any remains westward of the present village. POCOCKE says, (p. 245.) "Near the west corner of the bay was the port and town of Cysamus—the port was a small basin within the land, which is now almost filled up; it was defended from the north winds by a pier made of large loose stones, not laid in any order. Along the shore to the west of the port of Chysamo, there are foundations of some considerable buildings, which might be warehouses; a small rivulet runs into the sea at this port; and east of it the ancient Cysamus seems to have stood." After mentioning the castle

inquiring about it, I learnt that a sheep had fallen sick, and had thus been killed "for the children at Easter[36]." Not wishing, therefore, to rob the children of any portion of such an Easter Sunday's dinner, we supped on eggs and a salad of wild asparagus, which, if not quite so delicious as, like that met with in some parts of Italy, to compensate any epicure for the pains of a pilgrimage to eat it, was still most palatable after a ride of several hours. Belon observed, that asparagus, such as we cultivate in our gardens, could scarcely be seen in Crete, and that the wild plant grew all over the island[37].

The ancient epicures were well aware of the superiority of this wild vegetable over the larger asparagus which is produced by cultivation[38].

castle and the walls of the modern town, he adds, "as they are so near the sea, they would not be secure from the corsairs without this defence."

[36] *Διὰ τὰ παιδιά—νὰ φάγωσι τὴν Λαμπρή.*

[37] BELON, Observations de plusieurs Singularitez, etc. f. 20. ed. 1555. "A peine pourroit on voir celle maniere d'Asparges en Crete, telle que nous cultiuons en noz iardins: car ils n'ont que la sauuage nommée Corruda, qui y croist en tous lieux. Mais outre ceste là, ils en ont encor vne autre espece qui de nom propre vulgaire et ancien est appellée Polytricha." Wild asparagus is also found on the continent of ancient Greece: see CHANDLER, Travels in Greece, c. LI. near the end.

[38] ATHENAEUS, II. 62. e. *Οὗτοι καὶ ἕλειοι καὶ ὄρειοι καλοῦνται· ὧν οἱ κάλλιστοι οὐ σπείρονται, πάντων ὄντες τῶν ἐντὸς θεραπευτικοί· οἱ δὲ σπαρτοὶ καὶ σφόδρα ὑπερμεγέθεις γίνονται.*

WALLS OF POLYRRHENIA.

CHAPTER XXIV.

SITE AND RUINS OF POLYRRHENIA. RIDE TO MESOGHIA. THE ISLAND SAEGILIA, THE CERIGOTTO OF MODERN EUROPEANS AND THE AEGILIA OF THE ANCIENT GREEKS. ANCIENT AND MODERN CRETAN WINES. AKTE THE SITE OF KALE AKTE. THE PROBABLE SITUATION OF ACHAEA. THE STAGS OF CRETE EXISTED NOT IN THE TERRITORY OF ACHAEA, BUT IN THAT OF CYDONIA. ARRIVAL AT THE HAMLET OF KAVUSI.

April 23.

THE clouds looked very threatening this morning, and I had scarcely commenced my ascent to the ancient acropolis, before I was incommoded by a drizzling rain. In Crete rain usually falls on four or five days during the month of April, but never sufficiently for the people to speak, as our Chaucer does, of the time

"Whanne that April with his shoures sote
The droughte of Marche hath perced to the rote."

Besides, the drought of March is a thing sufficiently rare here, since I believe that month to be ordinarily one of the most rainy in the year. At all events the Cretans are anxious that it should be so, and shew, by a common proverb[1], their opinion, that the more rain they have in March the better, while with us in England dry weather is so much desired at the same season, that, according to our proverb, "A peck of March dust is worth a king's ransom." Not only in England, but even in Spain, April, and not March, is the month when the influence of genial and abundant showers is most felt[2].

In spite of the showers I examined the remains. From the summit of the acropolis Kísamo-Kastéli is south-south-west, and is about two miles distant in a right line. The ríver of Palaeókastro, which is also the river of Kamára, passes close to the east of the acropolis. From the western corner of the citadel, the walls run in a westerly direction for about three hundred paces, and their height varies from ten to eighteen feet. The stones, which are generally between two and three feet long, and from ten to twenty inches high, are disposed in horizontal layers. I should think there can be no doubt that these are remains of the walls built by the Achaeans and Laconians, when they came and settled with the Polyrrhenians and fortified this strong place: and I have thought it worth while to give a sketch of them. Somewhat to the south of this wall, and about half way between the acropolis and its western extremity, are remains of what I suppose to have been a temple. The modern Greek chapel which has been erected on its site is now also in ruins. On the acropolis itself but

[1] The proverb, which I heard every rainy day through the Greek month of March, is *Μάρτιος βρέχει, ποτὲ μυφιάζει*, where the *μυφιάζει* must be a corruption of *μυδιάζει*, as θ, instead of δ, is sometimes changed into φ, in the continent of Greece, and as conversely the Cretan village Viáno (*Βιάνο*) is sometimes called Dhiáno, (*Διάνο*.) *Μυδιάζω* I suppose to be derived from the ancient *μυδιᾷν*, which is well explained by Ruhnken, on Timaeus. See also Photius, Lex. p. 278: *Μυδιᾷν, νοτιᾷν, διυγρὸν εἶναι, καὶ σήπεσθαι.*

[2] "April showers, May flowers:" "April, aguas mil."

few vestiges of ancient walls are discoverable: there are however sufficient remains of them to shew that the Saracens, Byzantine Emperors, or, more probably, the Genoese, in their superstructure, followed generally the ancient foundations. Within these walls are some cisterns[3]. The circumference of the acropolis is about half a mile: but the city also was surrounded by walls, a portion of which we saw before arriving at the village. On the ancient site, at no great distance from the present village, the rock, a pebbly compound, is hewn in such a way as plainly to shew that its excavations once served as houses in the city of Polyrrhenia. To the westward of the village is the ruined tower before mentioned. It is about forty feet high, and is composed of stones of every size and of marble fragments, which indicate, as the period of its construction, an age when ancient monuments were regarded as useful only for building materials. Hard by this tower is a fountain, which is in fact the mouth of an aqueduct, about four feet in height and two wide, hewn out of the living rock, and said to go *an hour* underground, an assertion, however, which I did not think it worth while to verify by exploring its interior. At some little distance from the tower are also seen several ancient sepulchres.

Polyrrhenia, or Polyrhenia, as Spanheim, following the epigraphs of coins[4], would write the word, was the

[3] See POCOCKE, Vol. II. Part I. p. 246. BELON, in the first book of his Singularitez, Ch. V. fol. 7. describes the site. "A demie lieue de Chysamo tirant uers Cauo spata, ou Capo spada, lon trouue les ruines d'une ancienne ville sur vne colline à demy mile de la mer, ou encor sont restées les vestiges des murailles, et si grande quantité de belles cisternes, qu'il n'y a celui qui les puisse contempler sinon par grand miracle: les habitãts la nomment Paleo Helenico castro." In Ch. LVII, when describing the remains of Bucephalus at Cavallo, fol. 58. he again speaks of the cisterns of this Cretan Palaeókastron, "un peu au delà de Quissamus."

[4] Most of the coins have ΠΟΛΥΡΗΝΙΩΝ. STEPHANUS of BYZANTIUM says: Πολύρην, πόλις Κρήτης, ἀπὸ τοῦ πολλὰ ῥήνεα, τουτέστι πρόβατα ἔχειν. ὁ πολίτης Πολυῤῥήνιος. PLINY, IV. 20. calls the place Polyrrhenium. PTOLEMY, Πολυῤῥηνία; and SCYLAX, Πολύῤῥηνα; which should probably be either the Πολυῤῥηνία of Ptolemy, or the Πολύρην of Stephanus.

neighbouring state to Cydonia[5], and we learn from Scylax[6] that its territory extended from the northern to the southern sea. It was here that Agamemnon, when driven into Crete, came and offered sacrifice[7]. The celebrated temple of Dictynna was in its district[8], and it appears to have been a place of peculiar sanctity in ancient times[9]. Its colonization by Achaeans and Laconians has been already mentioned[10]. The Polyrrhenians took a part in the wars against Cnossos and Gortyna, which are related by Polybius[11].

We have seen[12] that, ever since the revival of letters, this Palaeókastron, in the neighbourhood of the western Kisamos, has been supposed to be the site of Aptera. In addition to the authors already mentioned I might quote Andrea Cornaro, the Cretan historian of his native island, and Ferrarius[13].

Among the extensive ruins of this city, and near one of its gates, Tournefort observed, inscribed on a long piece of stone, IMP. CAESAR.; and, on another fragment, IVII. COS. III[14]. After mentioning this he adds, "Tout cela marque que la ville a été considérable dans son temps, et il n'y auroit aucun doute que Paleocastro ne fust le reste de l'ancienne ville d'Aptére, n'étoit que Strabon ne la place qu'à dix milles de la Canée." The naturalist's way of getting rid of this difficulty is amusing: "Mais il n'y a rien de bien certain touchant les mésures des anciens, ou

5 See the passage of Strabo, translated at full length in Vol. I. p. 48.

6 SCYLAX, p. 18. ed. Huds. or Tom. I. p. 264. of the Geograph. Graec. Min. ed. Gail.

7 ZENOBIUS, v. 50. SUIDAS, in Οἱ Κρῆτες, and Professor HOECK, Kreta, Vol. I. p. 27.

8 STRABO, X. p. 479. Πολυῤῥήνιοι· παρ' οἷς ἐστὶ τὸ τῆς Δικτύννης ἱερόν.

9 SUIDAS, Πολύῤῥηνον, τόπος Κρήτης ἔνθα τοῖς θεοῖς ἔθυον.

10 See above, Vol. I. p. 48.

11 POLYBIUS, IV. 55.

12 See above, Vol. I. p. 54.

13 ANDREA CORNARO, ap. Cornel. Cret. Sacr. I. p. 123. and FERRARIUS, Lexic. Geograph. v. Cisamus, p. 101. ed. Lond. 1657.

14 POCOCKE, (Inscript. Antiq. P. I. c. 4. p. 43.) copied a Greek inscription here: see above, Vol. I. p. 41.

peut-être que cet endroit de Strabon est corrumpu[15]." Pococke did try to extricate himself from the difficulty, which these ten miles between Cydonia and Aptera forced on him; and in order to do so supposed, as we have already seen, that Cydonia was an inland city, situated five miles to the south-south-west of the modern Khaniá. Probably it is to this difficulty of reconciling the distance of Strabo with a false hypothesis, that we are also indebted for the indication of the imaginary ruins of Cydonia as existing at Ierámi. It is worth while to remark, that as Tournefort's description of this site agrees well with Strabo's account of Polyrrhenia, so Pococke observes that "the town was very strong by nature."

We left Palaeókastro at three; and, after a descent of about a mile and an ascent of another, passed the first of three or four hamlets, which are known under the common name of Lusákies. At half past four we arrived at the principal village of Mesóghia, where we are about two miles from the sea. Hence Tserígo and the islet called Cerigotto by the Italians, are both distinctly visible. The latter is called Saeghiliá by the Greeks, who have only prefixed an S to its ancient name[16], as has also been done in Sitanos, the name of

[15] TOURNEFORT, Voyage de Levant, Tom. I. p. 81. D'ANVILLE, Oevres, (Géographie ancienne abrégée,) Tom. II. p. 190. ed. Paris, 1834, like every body else, considers this Kísamo near Cape Spádha, as the port of Aptera.

[16] This island Aegilia is called Aegila by DIONYSIUS PERIEGETES, v. 499. Αἴγιλά τ' ἠδὲ Κύθηρα Καλαύριά τε τρηχεῖα· on which passage Eustathius shews pretty plainly that he knew nothing about its situation. It is mentioned by POMPONIUS MELA, II. 7. 11. and by PLINY, IV. 19. Aegila autem XV. M. pass. a Cythera, eademque a Cretae Phalasarna oppido XXV. M. pass. STEPHANUS BYZANT. Αἰγιλία—ἔστι καὶ νῆσος μεταξὺ Κρήτης καὶ Πελοποννήσου Αἰγιάλεια. ὁ οἰκήτωρ Αἰγίλιος ὡς Κυθήριος. Modern travellers as well as geographers (as D'ANVILLE, Geographie ancienne abrégé, Oeuvres, Tom. II. p. 191. MANNERT, Geographie der Griechen und Roemer, Vol. VIII. p. 690. CRAMER, Vol. III. p. 200. See also TZSCHUCKE, on Pomponius Mela, Vol. III. Part. III. p. 740.) describe the island by its Italian name Cerigotto, which is still as totally unknown to all the Greeks here, as Zante is for Zákythos, or Zákynthos (ἡ Ζάκυνθος) Gianitzares for Dionysiádhes (αἱ Διονυσιάδες), Stalimene for Lemnos (ἡ Λῆμνος), or Candia for Crete.

a modern village close by the site of Itanos, one of the easternmost cities of ancient Crete. A similar change is said to have taken place in the name of the Messenian city Andania, the site of which is now called Sandáni[17]: and thus Cardamyle, in Laconia, has become Scardamúla[18]: and, with the Italians, Chios is now Scio, Carpathos Scarpanto, and so forth.

The only Mohammedans of Mesóghia are the two Súbashis[19]. A considerable quantity of wine is produced here, and though it is not esteemed equally with that of Hághio Mýro, Sárko and a few other places, yet it is excellent; and, the produce exceeding greatly the home consumption, it becomes the main article exported from the district. The Cretan wine is frequently spoken of by ancient authors[20]. Hence our learned poet mentions it as one of the chief wines of antiquity[21]. Vines peculiar to the island are also spoken of[22]; and

17 GELL's Itinerary of the Morea, p. 69. The change of *Ios* into *Nio* may be noticed as analogous, the last letter of a different case of the article having there been taken and prefixed to the name; compare what has been said above, Vol. I. p. 264. An opposite change, the *omission* of the S, is found in the name of India. The river Sindhu in Sanskrit, became Hindu in Persian, and the aspirate was omitted by the Greeks, in the name of both the river and the people. Ἰνδὸς ποταμὸς ἀφ' οὗ Ἰνδοί. See BOHLEN, Das alte Indien, Vol. I. p. 9. and SCHLEGEL, De l'Origine des Hindous, Transactions of the R. S. L. Vol. II. p. 405.

18 LEAKE, Travels in the Moréa, Vol. I. p. 331.

19 Οἱ δύο Σουμπάσιδες.

20 CLEMENS ALEXANDRINUS, Paed. II. p. 185. ed. Pott. Ὁ εὐώδης καὶ εὔπνους Λέσβιος, καὶ Κρὴς τις γλυκύς. AELIAN, Var. Hist. XII. 31. Γλυκύς τις καὶ Κρὴς ἄλλος. ATHENAEUS, X. p. 440. e. Τὸν δὲ καλούμενον πάσσον πίνουσι· τοῦτο δὲ ποιεῖται μὲν ἐκ τῆς ἀσταφίδος, καὶ ἔστι παραπλήσιος πινόμενος τῷ Αἰγοσθενίτῃ καὶ τῷ Κρητικῷ. DIOSCORIDES, and GALEN, cited by TOURNEFORT, Vol. I. p. 90. A passage of JUVENAL will be quoted below, at p. 74. PLINY, XIV. 9. "Passum a Cretico Cilicium probatur, et Africum, et in Italia finitimisque provinciis." SOLINUS, c. 11. "Larga vitis, mira soli indulgentia."

21 MILTON, Paradise Regained, IV. 117.

> Their wines of Setia, Cales, and Falerne,
> Chios, and Crete.

22 As the Θρινία of HESYCHIUS, where, however, a comparison of PAUSANIAS, I. p. 66. renders Τεθρινία the probable name. See HOECK, Vol. I. p. 35.

some wines derived their names from the districts where they were produced. The Thenaean wine is thus mentioned by Julius Pollux[23]. It is deserving of remark that the site of Thenae, as we have already seen, is very near the modern villages of Hághio Mýro and Sárko, both justly celebrated as two out of four or five places which produce the best wine in all Crete. It will be remembered that Thenae was near Cnossos[24]. Hence Nonnus describes a libation, poured out by the Cretan Asterios, on the funeral mound of his fellow-countryman Opheltes, as having consisted of

> A cup of sweet and fragrant Cnossian wine[25].

The Pramnian wine, spoken of by Homer[26], in all likelihood derived its appellation, in the first instance, from some place of Thrace or Asia, where it was produced[27]. In later times a Cretan wine received the name[28]. Belon supposes it to have been the same with the modern malmesey[29]; but it is pretty plain that the port of our days is the wine which most nearly resembles the ancient Pramnian[30]. The wine of Crete was so little like the Pramnian, that it is called *passum* by several ancient authors[31].

The growth of the vine here may perhaps have received a check, while the Saracens were masters of the island, in the ninth and tenth centuries; but, if so, it

23 POLLUX, VI. 2. According to the correction of Meursius, (Creta, p. 103.) for the old reading Θηραῖον.

24 Above, Vol. I. p. 224.

25 NONNUS, Dionys. XXXVII. 81.

Σὺν ἀχνυμένῳ δὲ Λυαίῳ
Ἀστέριος Δικταῖος, ὁμόγνιον αἷμα κομίζων,
Κνώσσιον ἀμφικύπελλον ἔχων δέπας ἡδέος οἴνου
εὔοδμον, δαπέδοιο χυτὴν ἐμέθυσσε κονίην.

26 HOMER, IL. XI. 639. Od. X. 235. ATHENAEUS, I. p. 10. a.

27 EUSTATHIUS, on IL. XI. p. 871. PLINY, N. H. XIV. 4.

28 AELIAN, l. c.

29 BELON, Observations etc. f. 21.

30 HENDERSON, History of ancient and modern Wines, p. 75.

31 BENTLEY has rendered it impossible for us to adopt Hardouin's explanation of the verse of Horace: "Cressa ne careat pulchra dies nota." On the *passum*, see CASAUBON, Animadv. in Athenaeum, p. 735.

soon recovered, and while Crete still belonged to the Byzantine empire, its sweet wine was again celebrated. Theodore Ptokhoprodromos mentions it, with that of Mytilene, as opposed to the Chian[32].

From the period of the Venetian conquest Italy again enjoyed Cretan wines, which were not long in finding their way into the other countries of Europe. At the moment of the great insurrection of the Venetian colonists in 1363, as detailed in several unpublished manuscripts of St Mark's Library, wine was one of the principal exports of the island[33].

Somewhat less than a century afterwards, Buondelmonti travelled in Crete, and wine still held the first place among the exports[34]. About the same time Prince Henry of Portugal sent to Crete for plants to stock the island of Madeira, where the first Portuguese colony was established in 1421[35].

The wine of Crete is said by Aeneas Sylvius, who also flourished in the fifteenth century, to have been in

[32] THEODOROS PTOKHOPRODROMOS, II. 351.

Οὐδὲ γὰρ οἶνον πίνουσιν Χιώτικον εἰς κόρον,—
ἀλλὰ γλυκὺν Μιτυληναῖον καὶ Κρητικὸν ἀθύρην,
ἵνα χυμοὺς ἐκβάλλωσιν τοὺς τῆς ξηροφαγίας.

[33] CRONACA VENEZIANA, Codice XVIII. (belonging to the early part of the fifteenth century: see the Catalogo della Biblioteca Marciana, p. 226.) fol. 96. "Anchora li Centillomeni de Chandia si fexe elli uno p' so ducha e retore, de la tera—e fexe retegnir le nauue e nauuilli da Veniesia, ço i paroni e nocheri e marineri *le qual nave se trovaesse charegade alle vendeme.* In another manuscript, in the same collection, entitled also CRONACA VENEZIANA, and numbered Codice XX. in the Catalogo della Biblioteca Marciana, p. 227. we find at fol. 87. the same fact detailed, and the ships spoken of as "*tute le nave da vendeme andade ī Candia.*"

[34] BUONDELMONTI, in Cornel. Cret. Sacr. Vol. I. p. 9. end. "Veniunt ex omnibus mundi partibus huc naves, quae XX millia ad minus onerantur vegetibus optimi vini, et pinguis casei copia, ac frumenti, propter tamen rusticorum inertiam olivas non habent." Again, at p. 96. the same thing is told in nearly the same words. Towards the end of the sixteenth century, 60,000 botte was the amount exported annually: see FOSCARINI, Relatione, etc.

[35] ALVISE DA MOSTO, in RAMUSIO, Delle Navigationi et Viaggi, Vol. I. fol. 98. "Fece mettere piante ouero rasoli di maluasie, che mandò a torre in Candia, quali riuscirono molto bene." This passage is quoted by HENDERSON, p. 248.

great request even in Bohemia[36], and a Carthusian monk, who visited the island in 1507, on his way to the Holy Land, makes especial mention of the Cretan wine and honey[37]. The commerce between Crete and England, during the fifteenth and sixteenth centuries, was so great that Henry the Eighth, in 1522, appointed one Balthazari as "the Master, Governor, Protector and *Consul* of all and singular the merchants and others his lieges and subjects within the port, island and country of Crete or Candia[38]. The staple export of Crete was its wine[39], and the return obtained by it from England consisted chiefly of woollen cloths[40], a branch of commerce which was subsequently obtained by the French. Thevet[41] and Belon[42] both write in the middle of the sixteenth century, and both bear testimony to the abundance and excellence of the Cretan wines. An English traveller,

36 Aeneas Sylvius, Hist. Bohem. Cap. i. in Cornel. Cret. Sacr. Vol. i. p. 56. "Quoties Cretense vinum caupones venale proponunt, invenies quamplures, qui juramento adacti, nunquam cellam vinariam egredientur nisi exhausto dolio."

37 Cornelius, Cret. Sacr. Vol. ii. p. 416.

38 Rymer's Foedera, Vol. xvi. p. 766. The document, relating as it does to the first known appointment of a modern Consul, is interesting. The following is an extract: "De consule constituto. Dedimus et concessimus—dilecto et fideli nostro Censio de Balthazari, filio quondam Johannis de Balthazari, alias dicto Censio de Menesava, mercatori de Luca—modo residenti in insula Cretae vel Candiae—officium sive locum Magistri, Gubernatoris, Protectoris sive Consulis omnium et singulorum Mercatorum, aliorumque Legeorum et subditorum nostrorum, intra portum, insulam sive terram Cretae vel Candiae frequentantium, morantium, mercandisantium sive negotiantium."

39 The malmesey and muscadine, spoken of by our early poets, and in Shakespeare, were imported from Crete, which must have been, to our ancestors, what Portugal has been since. The Duke of Clarence drowned himself in his butt of malmesey in 1478.

40 Anderson's Origin of Commerce, Vol. ii. p. 60. where he speaks of 1534 and of 1550.

41 Thevet, Cosmographie de Levant, fol. 23. "Il y ha—grand' abondance de tresbons vins, cõme tesmoignēt les Epithetes poëtiq̃s et l'experience oculaire."

42 Belon, Observations etc. f. 21. "Le vin que nous appellons Maluaisie est seulement fait en Crete." Its European names were however derived from Monemvasía: Leake, Researches in Greece, p. 197. Moréa, Vol. i. p. 205.

in 1569, mentions the exportation of Malmesey as the common trade of the island[43]. The wood annually imported in order to make casks to hold it, was a considerable article of commerce. Sandys, who wrote more than forty years later[44], after speaking of the other produce of Crete, subjoins, "but that which principally enricheth the country is their Muscadines and Malmsies, wines that seldom come vnto vs vncuted[45], but excellent where not, as within the streights, and compared vnto Nectar.

> Creete I confesse Joues fortresse to be
> For Nectar onely is transferd from thee[46]."

The testimony of the English traveller is confirmed by that of a contemporary Italian Bishop, who was born in Crete, and who, when he speaks of his native land as vini ferax, immediately adds,

> Ignosce vini si excidit nomen mihi,
> Nectar volebam dicere, aut si quid magis
> Beat liquore lautà divum prandia[47].

[43] HAKLUYT's Collection of early Voyages, Travels, and Discoveries of the English Nation, Vol. II. p. 230. ed. Lond. 1810.

[44] SANDYS, a relation of a Journey begun An. Dom. 1610. containing a description of the Turkish Empire, of Aegypt, of the Holy Land, of the remote parts of Italy, and Ilands adioyning. The third edition. London, 1632. The extract is from p. 224.

[45] This word will be explained by BELON, f. 21. "Osons asseurer que celuy qui est trãsporté le plus loing, comme en Almagne, France, Angleterre, a éste premierement *cuict:* Car les nauires qui abordent en Crete pour trãsporter la Maluaisie en estrange pays, se veulent expressement charger de celle de Rethymo, sachants bien qu'elle se garde moult long temps en sa bonté, et que d'autant qu'elle est plus trauaillée, elle est d'autant plus excellente." He differs from the English traveller in the assertion of his last sentence.

[46] Vera equidem fateor Jovis incunabula magni:
Nam liquor haud alibi nectaris ille venit. J. C. SCALIGER.

[47] ALOYSIUS LOLLINUS, "clarissimus Praesul et Praesulum decus," in his verses entitled de Creta Insula, in Cornel. Cret. Sacr. Vol. II. p. 442. He was Bishop of Belluno for many years, and among other "egregia ingenii sui monumenta" were many writings "quae in Cretensium laudem cadunt." See Cornel. l. c.

Another testimony to the excellence of the Cretan wine is also afforded by a passage of Ben Jonson[48]. In "The Generall Historie of the Turkes," by Knolles, published at London in 1603[49], I find the island spoken of as "now most famous through a great part of the world, for the good Malmesey which there groweth, and is from thence in great abundance sent into many farre countries." England is mentioned as one of these countries by the Proveditor General, Foscarini[50], in his report to the Senate at Venice, made in 1576; and he also speaks of the excellence of the Cretan wine. Wine seems to have been produced in great quantities in the island, till it came into the possession of the Turks. Since that event the juice of the Cretan grape is rarely met with out of the island, but all modern travellers, who have tasted it, are unanimous in celebrating its praises[51]. I must quote the very words in which they are sung by Falconer.

Relaxed from toil the Sailors range the shore,
Where famine, war, and storm are felt no more;
The hour to social pleasure they resign,
And black remembrance drown *in generous wine*[52].

The whole district about Mesóghia is divided into vineyards. The three little islands seen off the western coast, were once, as I am told by the villagers, the

[48] BEN JONSON, The Fox, Act I. Sc. I. Vol. III. p. 173. ed. Gifford.

Like the rich merchant, who has filled his vaults
With Romagnia, or rich Candian wine,
Yet drinks the lees of Lombard's vinegar.

[49] It is a folio volume: the extract is from p. 868.

[50] FOSCARINI, Relatione etc. fol. 129. He says about 60,000 botte were produced annually. "Ne navigano molti per Ponente, parte in Inghilterra, parte in Portogallo per India, (they were chiefly muscadines, moscatelli,) molti ne vanno in Costantinopoli et nel mar maggiore, et assai in Alessandria dove non si consumano altri vini che di Candia."

[51] Tournefort, Louis Chevalier, Myller, and all other travellers in Crete down to TANCOIGNE, (Tom. II. p. 19.) might be referred to. From POCOCKE, Vol. II. Part I. p. 243. it appears, that, even when he visited the island, the Cretans exported wine "to all parts of the Levant," and that about Rhíthymnos they made "a fine muscadine wine."

[52] FALCONER's Shipwreck, Canto I.

castle of the Christians of these parts: most of their wives and families spent a spring and summer (from March to September) on them during the war. The places are called "the islands of the Akté[53]," Akté being the name of the district on the western coast hereabouts.

Now the fact that the word akté has gone out of use in the modern Greek language[54] converts this name into a vestige of antiquity, and enables us to identify the place, where it is found, with an ancient site. Kalé Akté is mentioned by Stephanus of Byzantium, on the authority of Eudoxus, as a city of Crete[55], but we know little more of it than its mere name.

The very meaning of the words Kalé Akté is a sufficient assurance that, although situated near the sea and on a beautiful coast, it did not possess a harbour. That an excellent port, on the south coast of the island, should have been called Fair Havens, was perhaps as natural as that this place should have had the name of Fair Shore. Nothing but the scantiness of our information respecting both the places, and the mistake which

[53] Τῆς Ἀκτῆς τὰ νησιά.

[54] The modern word is *γιαλὸς* or *γιαλί*, derived from the ancient *αἰγιαλός*. In Crete *γυρογιαλί* is very common. Not only have I never heard the word *ἀκτὴ* used by a Greek, but I do not find it in any modern Greek lexicon to which I have access, namely, the *Λέξικον τῆς Γραικικῆς Γαλλικῆς τε καὶ Ἰταλικῆς γλώσσης* of Spyridon Blante, published at Venice in 1816, and the *Θήσαυρος τῆς Ῥωμαϊκῆς καὶ τῆς Φραγκικῆς γλώσσας* (Paris, 1709.) of Alessio da Somavera, edited by Thomaso da Parigi, (both missionaries.) The words *γιαλός*, *ἀκρογιαλιά*, *ἀκροθαλασσιά*, and even *αἰγιαλὸς* itself, although it certainly cannot be said to be used by the common people, are all found. I have also searched in vain in the ATAKTA of Koray for any trace of the word *ἀκτή*, and therefore conclude that it is utterly lost to the modern language.

[55] STEPHANUS BYZANT. Καλὴ Ἀκτή, πόλις Κρητῶν. Εὔδοξος τετάρτῳ γῆς περιόδου· ἔστι δὲ ὡς μεγάλη κώμη. He adds that the gentile names of its inhabitants were Καλακτίτης, Καλοακτίτης, Καλακταῖος, and Καλοάκτιος. By taking what properly belonged to the Sicilian city, we may increase the number of these forms, from Suidas v. Καικίλιος by Καλακτιανός, and from coins (MIONNET, Tom. I. p. 221.) by Καλακτινος. Most of the old terminations of such words are still retained in the language. We have Ὑδραῖος, Καβουσιανός, Λακιώτης, Ῥιζίτης, &c.

I suppose to have been made by many commentators on a passage of the Acts[56], would make it necessary to observe that the occurrence of the same adjective (Fair) in their names, does not imply any identity between them. The place on the south coast, known as Fair Havens, was merely a port, and the adjacent town was called by a different name; while here, at Kalé Akté, it is only a city that is spoken of.

It would doubtless have been satisfactory to have found the old adjective still prefixed to the name of this place. But it is not surprising that it should have been lost. The capital of the island is now called, indifferently, the Great Kástron (Megálo-Kástron) and the Kástron. If it had ever been known by the former name alone, we should still easily recognize it by the latter. I therefore conclude that a part of the fertile district of Mesóghia, belonged in ancient times to the little city of Kalé Akté. Its domain would doubtless be bounded on the north by the Phalasarnian, and on the west and south by the Polyrrhenian territory. In all probability Kalé Akté itself belonged to Polyrrhenia. To suppose it independent, when in the immediate neighbourhood of states so much more powerful, seems to be out of the question.

It may be added that the name Kalé Akté was not peculiar to this Cretan city. Places so called existed both in Sicily and in Euboea. The Sicilian city is well known to Numismatists. A part of the coast of Attica was also designated by this name Akté[57], and it was likewise applied to a considerable portion of the maritime region of Argolis[58].

The name of another Cretan city, Achaea, the exact position of which I have no means of determining, would lead us to suppose its origin connected with an Achaean colony; and the known establishment of Achaeans and

[56] Acts, xxvii. 8. [57] Strabo, ix. p. 391.

[58] See the passages of ancient authors, indicated in Mannert, Vol. viii. p. 667.

Laconians, at Polyrrhenia[59], renders it probable that it was situated somewhere in this part of the island. The Cretan city Achaea is said, by an old Scholiast[60], by the author of the Great Etymologicon, and by others, to have possessed a peculiar breed of stags. This assertion is erroneous, and arose from its author's having misunderstood the sense of a Greek word *ἀχαιίνης*, which denoted no particular breed of stags, but was appropriated to the animal when in its second year[61]. This has been shewn very clearly by several writers[62]. If we were unacquainted with the true meaning of the word in question, the old Scholiast's false etymology would perhaps lead us to place the city Achaea near Cydonia, on account of Pliny's assertion[63], that in Crete stags were found only in the Cydonian district; and would also make us consider Achaea "as a place remarkable for its peculiar breed of stags[64]."

The word thus used to denote a stag was also the name of a kind of cake, which was made for the festival of the Thesmophoria[65].

59 See above, pp. 47 and 49.

60 Scholiast on Apollonius Rhodius, iv. 174. *Ἀχαία ἐστι τῆς Κρήτης πόλις, ἐν ᾗ γίνονται ἀχαιϊνέαι λεγόμεναι ἔλαφοι, αἳ καὶ σπαθιναῖαι καλοῦνται.*

61 I found peculiar names, used by the Cretans of the present day, to denote a lamb or sheep according to its age: thus, *ἀρνὶ* being a lamb and *πρόβἀτον* a sheep, *ἕνα μάροπο* denotes the animal when about a year old. This I learnt in the district of Mirabéllo: probably the usage of *μάροπο* does not prevail throughout the whole island. In the Moréa the calf is called *μοσχάρι* during the first year, *παλαιομόσχι* in the second, *δαμάλι* in the third, and *βοΐδι* afterwards: see Colonel Leake's Travels in the Moréa, Vol. i. p. 20.

62 See Meursius, Creta, p. 17. and Hoeck, Kreta, Vol. i. pp. 430–431.

63 Pliny, N. H. viii. 58. "Mirabilius in eadem insula (Creta) cervos, praeterquam in Cydoniatarum regione non esse."

64 Cramer, Description of Ancient Greece, Vol. iii. p. 391.

65 Athenaeus, iii. p. 109. f. *Ἀχαιίναι. τούτου τοῦ ἄρτου μνημονεύει Σῆμος—λέγων ταῖς Θεσμοφόροις γίνεσθαι.* *Ἔλαφος* also was the name of a cake, *πλακοῦς ὁ τοῖς Ἐλαφηβολίοις ἀναπλασσόμενος*: Athenaeus, xiv. p. 646. e. See Lobeck, Aglaophamus, Vol. ii. p. 1064. (Epimetrum xiv.)

After a short stay at the chief village of Mesóghia, we rode on to the hamlet of Kavúsi[66], which is considered as belonging to the same little district, and is near the shore. It is the nearest inhabited place to Kutrí, as the site of Phalasarna is now called.

My host at Kavúsi was one of nine brothers, seven of whom fell by the sword during the revolution. Formerly this village contained ninety souls, but now there are only four families in it. Our supper consisted of artichokes, boiled and then served up swimming in oil, eggs and cheese: we had also an abundant supply of the wine of Mesóghia, which we found excellent.

66 Εἰς τὸ Καβοῦσι.

DISTANT VIEW OF GRABUSA.

CHAPTER XXV.

ISLANDS NEAR KUTRI. SITE OF PHALASARNA. A THRONE FORMED OUT OF THE SOLID ROCK, AND OTHER REMAINS OF ANTIQUITY. THE CITY CORYCOS. FRUITLESS EXCAVATIONS. JOURNEY FROM KAVUSI TO KAMPOSELORAKHO.

April 24.

THE unusual appearance of travellers in the hamlet draws together from the neighbourhood eight or ten individuals, all Christians, as might be expected in this district, soon after day-break. From them I become acquainted with interesting details of several of the melancholy events which happened, here and in the neighbourhood, during the war.

Petalídha[1], is the name of the northernmost of the three little islands, the second, opposite to which is

[1] 'Η Πεταλίδα.

Kavúsi, is called Megalonesí[2], in spite of its very moderate size, and the third Prasonesí[3]. My host, after telling me the names of these islands, exclaims, pointing out Tserígo and Saeghiliá, "Lo! those islands, which belong to the English, lie close to Crete, and still another has come from so great a distance as Misíri, and has become our master[4]." We have already seen that these three islands, two of which served as places of refuge to the Greeks during the late war, are called by the anonymous coast-describer[5], Jusagora, Mese, and Myle. This notice throws light, as Mannert has observed[6], on the two islands near Crete, called Mylae by Pliny[7].

It may be worth while, before visiting the existing remains of Phalasarna, to see what mention of it is made by ancient writers. The fullest notice of this port of Polyrrhenia, contained in any extant author, is that of Dicaearchus[8], who tells us, in a passage which seems to bear every mark of being a dilatation of what Scylax had written on the subject, (for several of the words used are the very same, and they occur in the same order,) that "Phalasarna is situated to the west, and possesses an artificial port, and a temple sacred to Artemis, and that the goddess is called Dictynna." The distance between Polyrrhenia and its port Phalasarna is given by Strabo as sixty stades[9]. The distances of Phalasarna

[2] Τὸ Μεγαλονησί.

[3] Τὸ Πρασονησί, so called, as my host told me, διατὶ εἶναι γεμάτο πράσα, because it is full of leeks. There is an islet of the same name near Mýcono: see TOURNEFORT, Voyage du Levant, Tom. I. p. 278.

[4] Καὶ μᾶς ἐκυρίευσε.

[5] STADIASMUS, in Geograph. Graec. Min. Tom. II. p. 497.

[6] MANNERT, Geographie der Griechen und Roemer, Tom. VIII. p. 690.

[7] PLINY, IV. 12. §. 20.

[8] DICAEARCHUS, State of Greece, v. 118. (p. 24. ed. Huds.)

Φασὶ δ' ἐν Κρήτῃ πόλιν
εἶναι Φαλάσαρνα, κειμένην πρὸς ἥλιον
δύνοντα, κλειστὸν λιμέν' ἔχουσαν καὶ ἱερόν
Ἀρτέμιδος ἅγιον, καὶ καλεῖσθαι τὴν θεόν
Δίκτυναν.

[9] STRABO, X. p. 479.

from other places are mentioned by the Anonymous Coast-describer[10], who says that the city was sixty stades from the island Jusagora, and three stades from Mese; and adds that the third island was called Myle, and was fifty stades from the end of the promontory. These distances led Mannert[11] justly to conclude that the remains of Phalasarna were to be looked for to the north of Sfinári. Pliny[12] estimates the distance from the Malean promontory to Mount Cadistus at seventy-five miles: and Scylax[13] says, "It is a day's sail across from Lacedaemon to the promontory of Crete, on which is the city of Phalasarna;" adding that it is the first city to the west of the island.

At eight we set out to visit the remains of this city, and in an hour arrive at its site. The day is beautiful: in all probability yesterday morning's shower is the last we shall have for months. When once April draws to an end, scarcely either a single shower or a threatening cloud is seen to obscure the bright blue sky of Greece[14]. While in these highly favoured regions, where nature, at least, is still seen in all her pristine beauty, we wonder not that the ancient Cretans should have transferred to the bright day the very name of the Divinity[15], and that poets and philosophers should have identified the sky and atmosphere by which they were surrounded with the God of Heaven[16].

The plain running down to the shore contracts in width as we leave Kavúsi behind us: between Kavúsi

[10] In IRIARTE, p. 493. or in GAIL, Geogr. Gr. Min. l. c. (Tom. II. p. 479.)

[11] MANNERT, l. c. p. 690.

[12] PLINY, N. H. IV. 20. Item Cadisto a Malea Peloponnesi LXXV.

[13] SCYLAX, p. 17. ed. Huds.

[14] For many months the climate is so delightful, that we may well compare it to that of the quiet and happy abodes of the Gods, so beautifully described in the Odyssey, (VI. 183.)

[15] MACROBIUS, Saturnal. I. 15. *Cretenses* Δία τὴν ἡμέραν *vocant.*

[16] My reader will remember Chrysippus's doctrine, "aethera esse eum quem homines Jovem appellarent." (Cic. N. D. I. 15.) The sentiment is frequently brought forward by Euripides: see VALCKENAER's Diatribe, p. 47. foll.

and the sea, it is more than half cultivated; but, for a mile before arriving at the site of Phalasarna, all traces of human labour entirely disappear. The whole is said to have been cultivated before the revolution. The first object which reminds us that we are near the site of an ancient city, is a number of tombs hewn in the solid rock; there are near thirty of them: a little further is a great chair, which has also been cut out of the solid rock: the height of the arms above the seat is two feet eleven inches; and its other dimensions are in proportion. The chair, as well as the tombs which we have passed, is on the south-west side of the ancient city, the acropolis of which forms a conspicuous object from Kavúsi.

This enormous chair deserves some consideration, since the design of such a work is not very obvious.

We learn from Pausanias that many thrones were erected in Grecian temples, in honour of gods whose statues were generally in a sitting posture on them. Mr Quatremère de Quincy, in his great work on the Olympian Jupiter, has given an engraving of such thrones as may be supposed to have existed at Argos, Mantinea, Tegea, Sicyon, Patra and Megalopolis[17].

A throne was dedicated in the great temple at Olympia, by Arimnus king of the Tyrrhenians[18]; and, as we learn from Herodotus, Midas king of Phrygia, presented his own throne to the God at Delphi[19]. These were the first offerings made at Olympia and Delphi by any barbarian. Somewhat later the poet Pindar dedicated at Delphi a chair of less costly materials, but

[17] See PAUSANIAS, for Argos, II. p. 163. Mantinea, VIII. p. 616. Tegea. VIII. p. 695. Sicyon, II. p. 127. Patra, VII. p. 577. Megalopolis, VIII. p. 664. Of the last he says, Καθεζομένῳ δὲ τῷ Διῒ ἐν θρόνῳ παρεστήκασι, τῇ μὲν ἡ Μεγάλη Πόλις, ἐν ἀριστερᾷ δὲ Ἀρτέμιδος Σωτείρας ἄγαλμα. Mr QUATREMERE DE QUINCY has attempted the restoration of all these thrones in Pl. XIX. p. 322. of his splendid work, Le Jupiter Olympien. See the paragraph, "sur les trônes des divinités et autres monumens semblables dans les grands temples de l'antiquité," pp. 314—325.

[18] PAUSANIAS, V. pp. 405–406.

[19] HERODOTUS, I. 14.

THRONE OF PHALASARNA.

which was most highly valued, and had a place assigned it near the very shrine of the deity. It was made of iron, and its donor used to seat himself in it, whenever he visited Delphi, and to sing such of his hymns as celebrated the praises of Apollo[20]. The throne of Danaos was dedicated, in the temple of Apollo, by Lycios at Argos[21], and there was also a throne in the temple of the Mother of the Gods between the town of Corinth and the Acro-Corinthos[22].

It is unnecessary to dwell on works of ancient art so celebrated as the throne of Zeus at Olympia, or that of Apollo at Amyclae; but I will briefly mention such offerings of this kind as have come down to us, not merely in the descriptions of ancient authors, but in the hard and lasting materials of which they were made, or on which they were represented. In the temple of Themis at Rhamnus "a chair of white marble was placed on each side of the entrance; that on the right was dedicated to Nemesis, and the other to Themis[23]."

An ancient throne, curiously ornamented, has been found in the church of the Holy Virgin at the site of Chaeroneia[24]. Others are seen at Athens, at the Piraeus, to the north of Parnassus, and in the neighbourhood of Epidaurus[25].

[20] PAUSANIAS, X. p. 858. [21] PAUSANIAS, II. p. 153.

[22] PAUSANIAS, II. p. 121.

[23] UNEDITED ANTIQUITIES OF ATTICA, published by the Society of Dilettanti, pp. 51–52. The inscriptions on the two chairs are similar to one another. One is,

ΘΕΜΙΔΙ
ΣΩΣΤΡΑΤΟΣ
ΑΝΕΘΗΚΕΝ
ΕΠΙ ΙΕΡΕΙΑΣ ΦΙΛΟΣΤΡΑ

[24] The villagers call it the throne of Plutarch. See DODWELL, Tour through Greece, Vol. I. p. 32 and p. 222. It is spoken of by CLARKE, Travels, Vol. IV. p. 145. ed. 4to. and by LEAKE, Travels in Northern Greece, Vol. II. p. 115, from whom we learn that it is called ὁ φρόνος τοῦ Πλουτάρχου, as Θῆβαι is called Φῆβα.

[25] DODWELL, l. c.

Among the numberless treasures brought to light by the excavations at Ercolano are seen empty thrones of Mars and Venus[26]. They are both veiled or covered, as is likewise a throne of Saturn, represented on a bas-relief, which has been engraved and described by Mr Millin[27]. Two others, made of white marble, in excellent preservation, and highly ornamented with sculptured emblems which refer them to Bacchus and Ceres, are now at Rome, in the galleries of the Vatican[28]. Thrones of Neptune[29] and Saturn[30] were found in the ruins of an ancient temple at Ravenna, nearly a century ago.

It appears from a passage of Porphyry, that in the cavern of Mount Ida, where Zeus had been fed by bees and watched by the Curetes, a throne was placed in honour of him[31].

[26] LE PITTURE D'ERCOLANO, Tom. I. Tav. XXIX.

[27] MILLIN, Monumens antiques inédits, Tom. I. Pl. XXIII. p. 218. Mr Millin says, "Sur le trône du dieu est un grand voile. Il est probablement allégorique, et désigne l'obscurité dont les temps sont couverts;" an explanation which a knowledge of the custom, which we see to have prevailed, of covering the thrones of other deities, requires us at once to reject. Moreover the opinion that Kronos was the same deity as Saturnus, and that they were symbolical of time even in the most ancient popular belief, seems to be wholly destitute of all solid foundation. It has lately found a supporter in BUTTMANN, Mythologus, Vol. II. p. 32. f.; but it is sufficiently refuted by LOBECK, De Dodecatheo, Part II. pp. 5. and 10.

[28] VISCONTI, Museo Pio-Clementino, Tom. VII. Tav. XLIV. Trono di Bacco, and Tav. XLV. Trono di Cerere.

[29] BELGRADO, Il trono di Nettuno illustrato, Cesen. 1776. MONTFAUCON, Antiquité expliquée, Suppl. Tom. I. Ch. VII. Pl. XXVI. "Le trône paroit des plus magnifiques—un grand voile couvre le siége." QUATREMERE DE QUINCY, Le Jupiter Olympien, p. 315.

[30] MILLIN, Monumens antiques inédits, Tom. I. Pl. XXVI. Monumens du Musée Napoléon (Prioli), Tom. I. Pl. I. L'Abbé BELLEY, Mémoires de l'Académie des Inscriptions et Belles Lettres, Tom. XXI. pp. 421. and 427.

[31] PORPHYRY, Life of Pythagoras, p. 34. ed. Kiessling. *Εἰς δὲ τὸ Ἰδαῖον καλούμενον ἄντρον καταβὰς, ἔρια ἔχων μέλανα τὰς νενομισμένας τριττὰς ἐννέα ἡμέρας ἐκεῖ διέτριψε, καὶ καθήγισε τῷ Διΐ, τόν τε στορνύμενον αὐτῷ κατ' ἔτος θρόνον ἐθεάσατο.* This annual custom of covering the throne reminds me of the new garment which was made once a year for the Amyclaean Apollo. PAUSANIAS, III. p. 247. *Ὑφαίνουσι δὲ κατὰ ἔτος αἱ γυναῖκες τῷ Ἀπόλλωνι χιτῶνα.* The peplos of the Olympian Here

It must be allowed that the instances adduced do not completely illustrate the meaning of this enormous seat: but it is not surprising that the Greeks should have cut the natural rock into the form of an offering with which, as has been shewn, they were wont to honour their deities. We know how gladly they availed themselves of any physical objects which could be converted into monuments of their piety; and thus, as we have elsewhere seen[32], the sides of a cavern were covered with an address to its tutelary deity, or received the sculptured forms of the Nymphs who enjoyed its shade[33], while seats were hewn for them within its cool recesses[34].

I also find a similar work described, by Pococke[35], as existing near the site of Antaradus in Syria. He says, "There is a court cut into the rock, with a throne in the middle of it—the throne consists of four stones besides the plain pedestal, which is cut out of the rock—it was probably built for an idol worshipped in this

Here was also renewed every fifth year. PAUSANIAS, V. p. 417. Διὰ πέμπτου δὲ ὑφαίνουσιν ἔτους τῇ Ἥρᾳ πέπλον. See VI. p. 515. It appears from the Rosetta Inscription, that in Egypt the statues were clothed three times a day by the priests, that is as often as divine service was performed in the temple. L. 40. ΚΑΙ ΤΟΥΣ ΙΕΡΕΙΣ ΘΕΡΑΠΕΥΕΙΝ ΤΑΣ ΕΙΚΟΝΑΣ ΤΡΙΣ ΤΗΣ ΗΜΕΡΑΣ ΚΑΙ ΠΑΡΑΤΙΘΕΝΑΙ ΑΥΤΑΙΣ ΙΕΡΟΝ ΚΟΣΜΟΝ. The custom is also mentioned by HERODOTUS, II. 42. and II. 132. and PLUTARCH, de Isid. et Osir. p. 366. f. See DRUMANN, Untersuchungen &c. p. 106. foll. These ceremonies of the Egyptians and Greeks prevailed among the Romans, and have been duly adopted in the religion of modern Italy. As the temple of Baal, 2 KINGS, x. 22, possessed a wardrobe in former times, so the house of our Lady of Loretto is well provided with one now; and at Rome the great bronze statue of Saint Peter is splendidly habited on days of festival. (See MIDDLETON's Letter from Rome, p. 154. and DODWELL's Tour through Greece, Vol. I. p. 317.)

[32] At Melidhóni: see above, Vol. I. p. 139.

[33] As at Paros, where there is an inscription under the bas-relief: see TOURNEFORT, Voyage du Levant, Tom. I. p. 201. and above, Vol. I. p. 139.

[34] HOMER, Od. XII. 318.

Ἔνθα δ' ἔσαν Νυμφέων καλοὶ χοροὶ ἠδὲ θόωκοι.

VIRGIL, Aen. I. 167.

Intus aquae dulces, vivoque sedilia saxo,
Nympharum domus.

[35] POCOCKE, Description of the East, Vol. II. Part I. p. 203. (c. 27.)

court or open temple; and it is one of the greatest and most extraordinary pieces of antiquity that is to be seen[36]."

We learn moreover from Pausanias, that a throne was cut out of the living rock, as this is, at Temenothyrae, a small city of Lydia, where the common people called it the seat of the giant Geryones[37]. Probably its dimensions were not very different from those of the massive block before us.

Ptolemy Euergetes, after his extensive conquests in Asia, went down to Adule, a maritime city of Ethiopia, where he placed a sculptured throne, made out of a single block of white marble, on the spot which he had chosen for his seat while he reviewed the assemblage of his numerous armies[38]. Ptolemy's offering was dedi-

[36] BOETTIGER, Ideen zur Kunst-Mythologie, p. 244. observes, that this throne must have belonged to a national deity of Syria, and may very well have been designed for a statue of Kronos. Perhaps, however, no statue was ever placed on it, any more than in the seat of Geryones: the throne alone, as has been so fully shewn in the text, was an offering in honour of the deity.

[37] PAUSANIAS, I. p. 87. *Λυδίας τῆς ἄνω πόλις ἐστὶν οὐ μεγάλη Τημένου θύραι. ἐνταῦθα περιῤῥαγέντος λόφου διὰ χειμῶνα ὀστᾶ ἐφάνη—αὐτίκα δὲ λόγος ἦλθεν ἐς τοὺς πολλοὺς Γηρυόνου τοῦ Χρυσάορος εἶναι μὲν τὸν νεκρὸν εἶναι δὲ καὶ τὸν θρόνον· καὶ γὰρ θρόνος ἀνδρός ἐστιν ἐνειργασμένος ὄρους λιθώδει προβολῇ.* The city Temenothyrae was on the confines of Mysia, and became, in later times, a Christian see: vid. MANNERT's Geographie der Griechen und Roemer, Vol. VI. Part III. p. 115. Dr CRAMER, Description of Asia Minor, Vol. I. p. 60. mentions Temenothyrae as an obscure place noticed only by Ptolemy.

[38] COSMAS INDICOPLEUSTES, in CHISHULL's Antiquitates Asiaticae, p. 77. MONTFAUCON, Nova Collectio Patrum, p. 140. and VINCENT, Commerce and Navigation of the Ancients, Vol. II. p. 331. foll. Xerxes also seated himself in a throne, made of white marble, at Abydos: HERODOTUS, VII. 44. whose expression *προεπεποίητο γὰρ ἐπὶ κολωνοῦ ἐπίτηδες αὐτῷ ταύτῃ προεξέδρη λίθου λευκοῦ*, reminds me of the inscription, **ΠΟΤΑΜΩΝΟΣ ΤΩ ΛΕΣΒΩΝΑΚΤΟΣ: ΠΡΟΕΔΡΙΑ**, found on a throne of white marble, which I saw at Mytilene in 1833, and which is briefly mentioned, and well engraved, in POCOCKE, Vol. II. Part II. p. 15; and is also engraved in M. CHOISEUL-GOUFFIER's Voyage Pittoresque de la Grèce, Tom. III. Part. I. Pl. 8. p. 85. A silver-footed throne served the Persian monarch as a seat during the battle of Salamis: see HARPOCRATION, in *ἀργυρόπους δίφρος*, AESCHYLUS, Pers. 466. and BLOMF. Gloss. v. 473. If any one should suppose the throne

cated to the God of War[39], and was placed at the very entrance of Adule[40], just as we find this monument within a few paces of the ruined walls of Phalasarna.

The maritime situation of the city might lead us to suppose the throne to have been dedicated, like that at Ravenna, to Poseidon; but the prevalence of Dictynna's worship in this part of the island, and the known existence of a temple of her's at Phalasarna, make it perhaps somewhat more probable that the offering was meant to honour that Cretan Goddess.

Remains of the city walls of Phalasarna exist in a greater or less degree of preservation, from its northern side, where it seems to have reached the sea, to its south-western point, cutting off the acropolis and the city along with it, as a small promontory. In the existing remains of these walls, near the sea on the north side, there are seen square projections, which we may suppose to have been the places of towers. One of these projections is found near the very northern extremity of the walls: it is succeeded by a curtain, if I may consider the towers as bastions, and may use a technical term of the art of fortification, one hundred and twenty feet long: another interval of about two hundred and thirty feet brings us to a third tower, the face of which is thirty-six feet, while its flanks are twenty feet long. A hundred paces more and we arrive at a little chapel dedicated to Hághios Gheórghios; and one hundred and twenty feet further on is another projecting tower or bastion, after passing which the direction of the walls changes, and, instead

throne at Phalasarna to have been thus meant as a seat of honour for men, rather than as an offering to a deity, he may compare SUIDAS, under *οἱ νομοφύλακες*, VALESIUS, on Harpocrat. 55. and WALPOLE, Memoirs relating to European and Asiatic Turkey, Vol. I. p. 310. The throne represented on the coins of Olba in Cilicia, and on which there is a dissertation by the Abbé BELLEY, in the Mémoires de l'Académie des Inscriptions et Belles Lettres, Tom. XXI. pp. 421 and 427. may either have been dedicated to Zeus, or used by Polemo.

39 CHISHULL, p. 78. or pp. 81–82.

40 COSMAS, in CHISHULL, p. 74. Ἐν τῇ ἀρχῇ τῆς πόλεως κατὰ τὸ δυτικὸν μέρος, προσέχων ἐπὶ τὴν ὁδὸν Ἀξιώμεως.

of proceeding southward, turns to the east. Following them in this new direction for about one hundred and twenty feet, on an elevated ridge where their foundations are visible all the way, we arrive at a mound of large hewn stones, lying about in such confusion, that it is impossible to say of what building they once formed a part. One naturally thinks of Dictynna's temple. From this point the course of the walls again changes, and resumes very nearly its original southerly direction. Their length from this point to the southern sea, near the artificial port spoken of by the ancient writers, and the situation of which is immediately discerned, is about two hundred paces.

The towers found in some ancient walls resemble still more nearly the modern bastion than those seen here at Phalasarna. The nearest approach of the kind, which I have noticed, is in the ruins of Priapus, the celebrated city on the Hellespont. Remains of walls, built probably at an early period after the transfer of the seat of empire to Constantinople, still exist there, with towers at short intervals, which project beyond the line of the wall, and present something very like the requisites of a bastion, two flanks, each at right angles to the wall, and two faces, each inclined to the flanks and to each other at given angles[41]. Here we have only the simple

[41] I shall be pardoned for making the following extract from my journal, since I believe these remains of Priapus have not been described by any modern traveller: "Sunday, 22d September, 1833. The city of Priapus occupied the whole of this narrow tongue of land, the shape of which may possibly have led to its selection as a suitable place for the worship of its tutelary deity, and may have bestowed its name on the city. The whole of it appears to have been surrounded by walls, which however were not very strong, except on the land side, and in fact, with such a coast, were but little needed elsewhere. Across the neck of this promontory the walls extend from sea to sea, and have regular towers at very short distances from one another. There are distinct and considerable ruins of ten of these towers, eight of which may be said to be well preserved. They are built in the common Roman style: the interior being made of bricks, loose stones of various shapes and sizes, and a great deal of mortar; while all this is covered on the outside with regular brick-work: a casing which is here in some parts three feet thick. I also find three cisterns for water, two of them communicating with one another, and

front parallel to the line of the walls, and its two flanks. But there are some other remarkable peculiarities about these walls: the first that they have a *slope*, as if constructed according to the rules given by writers on the modern art of fortification, and which every one will remember to be of invariable occurrence in both the bastions and curtains of regular fortifications, although it must be very rare, even if it ever occurs, in ancient cities: the second is that there is a double line of walls, and these towers or bastions are similarly situated in both. Very considerable remains of both the walls are seen as we pass along the first three hundred and fifty feet of their length on the north side of the city, and the distance between them is sixteen feet. As we follow their continuation from the confused heap of large hewn stones to the southern sea, it appears, both from the considerable width of the elevated ridge which indicates their position, and from some still visible remains of them, that the system of double walls and towers was followed throughout their whole extent. The whole distance from sea to sea is about six hundred paces. The little chapel of Hághios Gheórghios is situated somewhat less than two hundred paces from the northern sea, and nearly four hundred from the south-eastern part of the city, where, close by the shore, are rocks in which stone-quarries have been worked, and this too with such regularity as to shew that the apartments thus

and each of them more than twenty-two feet long, and about fourteen feet wide and sixteen feet high. From these old towers we have a fine view of the Granicus, and of the whole of its plain, where the great battle between Alexander and Darius was fought, as well as of the mountains by which it is bounded. The river discharges itself into the sea right opposite to this Palaeókastron. The trees which surround the village of Dhemotikó, distant about twelve miles up the plain, form an object very distinctly visible. Near us we see the picturesque village of Karáboa, parts of its houses peeping out from among the trees which surround it, and with its mosque and minaret, the neighbouring windmills, and its beautiful cypresses, forming a picture which pleases the eye, and produces an illusion soon to be destroyed by rude realities, when we enter its precincts, and see the squalid poverty and wretchedness of its mud cottages."

formed, in the open air, were meant to be applied to some use. The principal entrance to the city appears to have been about fifty paces from the sea, near these excavations; and, between this entrance and the shore, a long and solid slip of the rock has been left standing, to serve no doubt as a continuation of the walls, which extended to this entrance from the other side of the city.

It only remains to describe the acropolis, which forms so conspicuous an object in the view seen from Kavúsi. After mounting up it I found, in a gap between two rocks, a small space, just before reaching which the ascent is aided by a flight of eight or ten steps. The interval between the highest step and the rock is not above eighteen feet long. The indications of an ancient building here are the shaft of a column and foundations of a wall. In ascending to this spot, which is about one hundred and ten paces from the northern end of the acropolis, and not far from the chapel of Hághios Gheórghios, the side of the hill is not quite so steep as it is elsewhere; and on this account it would seem that nine or ten walls which I passed have been built along its side. Their age is a little difficult to determine: at first sight, and from a distance, one would take them for a specimen of the most ancient of the so-called cyclopian styles; but, on a nearer inspection, they seem to bear a greater resemblance to those which are used in some parts of England as fences; and to have been put together without any very great care. I suppose the remarkable solidity of the construction to have been caused by the simple fact that these stones were found by the builders ready to their hand, and that the excess of labour in putting them together was compensated for by their not having to use the hammer or chisel. No doubt walls so put together might last two or three thousand years; but that these cannot pretend to any high antiquity is certain; for after passing, on descending, the nine or ten lines of them, and reaching the foundations of what were plainly the exterior walls of the ancient city, I see, on these still visible foundations, which are similar in appearance to the rest of the undoubted ancient masonry, a piece of walling of precisely the same character as those in question. Perhaps one may suppose this acropolis to have been made a stronghold, and, on this account, to have been thus rudely strengthened by the Greeks, when first attacked by the Moors in the ninth century; or when, early in the thirteenth century, they united themselves with Henry Count of Malta, supported by the Genoese, in attempting to make a stand against the Venetians, and to expel them from the island. But it may be alleged as an objection to the latter supposition, that all the contemporary fortresses are strengthened in a manner very different from this.

Walking along the acropolis to the south, I find remains of an ancient building on each of its summits. Whether they were forts or temples, the existing remains, than which I never saw a greater heap of confusion, do not enable me to say. From the south end of the citadel a sketch of the view to the north, including most of the rock and fortress of Grabúsa, was taken, and is exhibited at the head of this chapter. The promontory which we see is put down in Lapie as "Pointe Frizer," a name which, as is usually the case with the received names of capes, hills, rivers and even towns in maps of Crete, is unknown in the country: La poêle à frire would have been nearer the truth, for the Greeks call it the Fryingpan[42], on account of its shape. It was the scene of a conflict very disastrous to the Greeks, during their occupation of Grabúsa, sometime before they arose with Khadjí Mikháli.

The Grabúsian headland is called Kimaros by Strabo[43] and Tretos by the anonymous Coast-describer[44]: all other

[42] Τὸ Τηγάνι. The site of the ancient city Messa is on a promontory called Τηγάνι: see Colonel Leake's Morea, Vol. i. p. 286.

[43] Strabo, x. p. 474.

[44] In the Geograph. Graec. Min. Tom. ii. p. 497.

geographers call it Corycos. We learn too from Pliny[45] that the islands of Grabúsa, which lie off this promontory, were called Corycae. Mannert[46] and Dr Cramer[47] speak only of the promontory Corycos[48] and its mountain, and say nothing of a city of the same name which existed on it. Meletius mentions both the promontory and the city[49]; and I find in Golzius[50] a coin of the Corycians, the symbols on which are manifestly Cretan, being the same as are common on medals of the island, though perhaps this is the only independent single city in it on the coins of which they are found. To place the matter out of doubt, Ptolemy[51] speaks of Corycos as the name of the promontory *and city;* and there is a passage where Juvenal mentions a Corycian vessel which evidently belonged to this Cretan town[52].

Pococke[53] tells us that the Turkish garrison of Grabúsa "were such bad neighbours that the whole promontory was uninhabited;" and he could not hear of any remains, "there being only a small ruined convent of Saint George and two churches on the promontory." Remains, however, do exist, or at all events did exist when the Florentine traveller Buondelmonti visited Crete in the year 1415. He speaks of them as situated on a small plain at the summit of the hill,

[45] PLINY, N. H. IV. 20. "Reliquae circa eam ante Peloponnesum duae Corycae: totidem Mylae."

[46] MANNERT, Geographie der Griechen und Roemer, Vol. VIII. p. 691.

[47] CRAMER, Description of Ancient Greece, Vol. III. p. 365.

[48] Thus STEPHANUS OF BYZANTIUM, Καὶ ἀκρωτήριον Κρήτης Κωρυκίη, and PLINY, IV. 12.

[49] MELETIUS, Geograph. p. 409. Κώρυκος Ἄκρα, κοινῶς Κόρνικο, καὶ πόλις ποτὲ κατὰ τὸ δυτικὸν μέρος τῆς νήσου, ὅπου παράκεινται αἱ Κώρυκαι Νῆσοι, καὶ αἱ Μύλαι Νῆσοι.

[50] GOLZIUS, Graeciae et Asiae Minoris Nomismata, Tom. III. p. 221. (Tab. III. Insularum Nomismata,) ed. Antwerp, 1644.

[51] PTOLEMY, Geogr. III. p. 91. Κώρυκος ἄκρα καὶ πόλις. He is referred to by Dr Cramer, who makes no mention of the city.

[52] JUVENAL, XIV. 267.

Tu Corycia semper qui puppe moraris—
Qui gaudes pingue antiquae de litore Cretae
Passum, et municipes Jovis advexisse lagenas.

[53] POCOCKE, Travels &c. Vol. II. p. 246.

and as consisting of walls still of a considerable height, and remains of buildings, with cellars and cisterns excavated out of the living rock[54]. The words of this old Italian traveller "montem per immeabilem ascendo semitam," remind me of the expression of the Coast-describer[55], and of the accounts of the people at Kutrí and elsewhere, when speaking of the road to Grabúsa, which they all described as one of the very worst and the most difficult to pass in the island.

After spending several hours in the examination of the remains of Phalasarna, we obtained some misíthra, brought us by a shepherd whose sheepfold was near, and with it and some bread we appeased the hunger excited by our long morning's work, and began to retrace our steps to the hamlet of Kutrí. Our host Gheorgákes Ligopsýkhes[56], who, along with some other villagers, had accompanied us as a guide and companion, tells me of the discovery, some years ago, of certain "statues[57]" and "idols[58]," a little below the village. The Greek who found them, afraid of the fact's becoming known, and thus perhaps producing a suspicion that he had also discovered a hidden treasure, hastened to conceal them all: he mentioned his good fortune to no one till he was lying at the point of death, when he disclosed it, and described the place where he had concealed them. Captain Maniás, whose antiquarian

[54] CORNELIUS, Creta Sacra, I. p. 87. "Dum sic peragramur per longum iter aliquantisper sub radicibusque promontorii Coricis fessi recreamur, denique ad eundem montem per immeabilem ascendo semitam, in summitate cujus civitas cum plano exiguo eminebat, meniaque ejus non parum a terra elevata cernere poteramus. Mansiones insuper diligentissime fabricatas cum cellulis subterraneis clare videmus, et cisternas in lapide sculptas mire industrie collaudamus."

[55] ANONYMI STADIASMUS, p. 497. ed. Gail. Ἀκρωτήριόν ἐστι τετρημένον, κατάκρημνον τῆς Κρήτης.

[56] Γεωργάκης Λιγοψύχης, Georgy Littlesoul, if literally translated. The class of names derived from moral qualities is not very extensive either in modern Greek or in English. We have Good, Best, Goodenough, Sly, Coward, Coy, Gallant, Humble, Keen, Sharp, Wise, Kenworthy, Wordsworth, and others.

[57] ἀγάλματα.

[58] εἴδωλα.

zeal shews itself daily in the alacrity and indefatigableness with which he aids me in making inquiries for ancient remains, and for coins or other antiquities, is very anxious that I should not suffer all these hidden treasures to continue buried in the earth, and warmly seconds my host's wishes. At length, although the assertion that the "idols" are small and golden, makes me fear that it will end in nothing, for the form assumed by the story bears a slight resemblance to that of a buried treasure, which, of course, no European in his senses would ever search for; yet, the day being already very far advanced, I at length give way to the solicitations of the people, and agree to wait here till to-morrow at noon; the Greek undertaking to have the excavations finished by that time. Being unwilling, however, thus to stay, before I decide on doing so, I offer, but in vain, to give the man a written authority to excavate, which, on his own account, he would not have dared to do, and suggest that he may inform me afterwards of the result of his labours. His objection to doing any thing if I leave him is because the first finder died soon after making the discovery, and he believes my being present will avert from him a similar fate, which would assuredly await him, if he set to work after my departure.

April 25.

The excavations are commenced early this morning by half a dozen men, who continue their work six or seven hours without finding any thing. On this I became satisfied that nothing would be found, either within or outside the house, and, after recompensing the people for their trouble, I set off at three o'clock.

We soon commenced the ascent of the western slope of Mount Elías: the greater part of the olive-trees, amongst which our road lies, are uncultivated. This ascent lasts for forty minutes, and the descent by which it is succeeded is nearly as long, and at length brings us on a slope extending from the mountains to the shore, and along which we pass for between four and five miles, always at a distance of about two miles from the sea. Its continuity is broken by three small ravines: in the winter there flow down them torrents, which are already reduced to mere streamlets. At the third ravine, near the extremity of the slope, the village of Sfinári is on our left. Hence, after a steep ascent of nearly half an hour, we arrive at the summit of the ridge, which ends in what may be called Cape Sfinári, and, descending into a fertile and well-cultivated valley of corn-fields and olive-trees, we pursue it, in a southwardly direction, for about half an hour, when we arrive, at six o'clock, at the village of Kamposelórakhos[59]. We are still only about two miles from the sea, which is visible through another valley in a west-north-west direction.

Although we arrived before sun-set, the villagers, unable as they were to imagine what could be the object of our rambles among their mountains, were considerably alarmed on seeing so extraordinary a sight as European travellers. I should add, that in the houses near which we dismounted, no male inhabitant was at the moment to be found. One of the women asked me, What our business was, and why we had come there. I replied, laughing, "To see you to be sure." But her fears were too serious to be trifled with, and I could not at all quiet the suspicion with which she regarded us. We succeeded in finding a stable for our steeds, but since the good lady professed *not to have the key* of her house, we did not obtain a lodging for ourselves, but remained in the open air, till fortunately a man arrived in the village, and received us with the ready hospitality usually shewn. Our evening meal consisted of milk and eggs.

[59] Καμποσελόραχος.

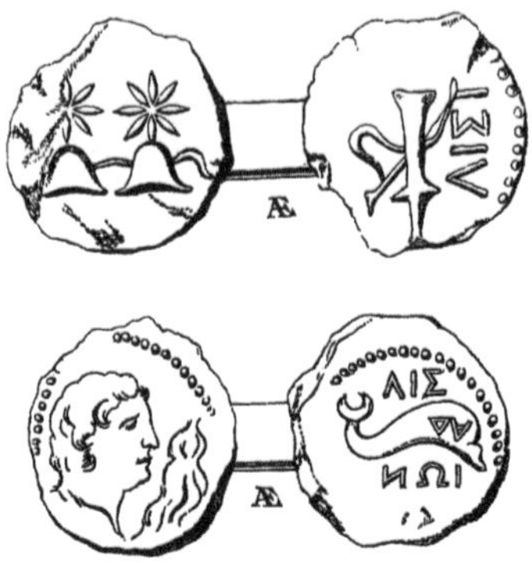

COINS OF LISSOS.

CHAPTER XXVI.

KAMPOSELORAKHO TO SKLAVOPULA. THE ANCIENT CITY DULOPOLIS. THE PRESENCE OF SLAVONIANS, IN THE CONTINENT OF GREECE, HAS NOT EXTENDED TO CRETE. SELINO-KASTELI. SFAKIAN DIALECT. SITE OF KALAMYDES. ANCIENT DORIAN INSTITUTIONS IN CRETE. HAGHIO KYRKOS, THE SITE OF LISSOS. TWO SUPPOSED COINS OF LISSOS. A NIGHT'S LODGING AT HAGHIO KYRKO.

Kamposelórakho, April 26.

THIS village now contains only forty houses: it pays fifty-eight kharatjs. Kamposelórakhos and Kunoné are the two principal villages of the district called Enneákhoriá. The similarity of sound between Inakhorion and Enneá-khoriá, reminds us of the ancient city which formerly existed on the sea-coast at no great distance from this place. I am inclined to believe the name of the district to be a corruption of that of the ancient city[1]. I do not, however, hear of any remains of an-

[1] Thus, I suppose, Hippocoronion has become Apokórona: see Vol. I. p. 63.

tiquity in the neighbourhood; and I am but little disposed to follow the shore from this place to Sélino-kastéli, a route which, in all probability, would require three days.

We have already seen that the territory of Polyrrhenia extended from the northern to the southern sea: hence, therefore, it is plain that the little towns on the coast between Phalasarna and Kriu-metopon, though recorded by many writers, and though each of them may have enjoyed a nominal independence, were yet so limited in their respective districts and resources, that their size and power must always have been very insignificant. And this want of physical greatness is not compensated for by the connexion of any of them with early religious or other myths, which might make them interesting. My reader will see them placed on the map either in or very near the places which they occupied; and, not having spent my own time in endeavouring to identify precisely the position of each of them with any ruins which may now exist and indicate their sites, a slight notice of them is all that will be wanted before we go on to the south coast. Ptolemy mentions Khersonesos, Port Rhamnus, and Inakhorion as existing between Phalasarna and Kriu-metopon. Not one of them is recorded in Pliny's list, though it contains forty Cretan towns. The Coast-describer[2] speaks of Bienon, which he fixes at twelve stades from Kriu-metopon: he mentions it as a place with a port, and where there was water, an indication which he frequently makes, and which of course was as useful a direction for sea-faring people in ancient times as it is now a-days.

We leave Kamposelórakhos at eight. A descent of a few minutes brings us to a cascade, the first which I have seen in the island: the fall, of not more than twenty feet, is over rocks into a streamlet. We ride

[2] STADIASMUS, p. 497. ed. Gail. (Tom. II. Geograph. Graec. Min.) Ἀπὸ Κριοῦ Μετώπου περίπλους εἰς Βίενον σταδ. ιβ'. λιμένα ἔχει καὶ ὕδωρ.

along a slope running gradually down to the sea, which is distant only about a mile. The mountains are covered hereabouts with heath, which is now in flower, and wild strawberry trees[3]. The fruit of this shrub is gathered in September and October for making rakí: I am also assured that it is pleasant to the taste; which is certainly more than can be said of the spirit extracted from it. In the silk districts, mulberries, which the Greek believes to be unwholesome and can hardly be persuaded to taste, are either given to pigs or converted into a rakí, perhaps the most execrable of all the preparations that bear the name. We pass the hamlet of Keramúti[4], and, a few minutes after nine, arrive at Amýgdhalokepháli[5]: our road continues to be in the same direction, and, after passing the villages of Makerianá[6] and Vavulianá[7], we arrive, at half past ten, at Kunoné[8]. We had overtaken a Christian peasant of this place a mile before arriving at it, and I, as usual, immediately began to talk with him, and continued doing so till we arrived at the village. As we approached it I saw an Arnaut, whom, in Crete, one can at once recognize by his dress, so different from the Cretan costume, as well as by the musket slung over his shoulder and the pistols hanging in his girdle. I observed to the Greek, "So you have got some Arnauts in your village; how many are there?" He made no reply. I repeated my question, and the poor fellow, evidently in great alarm, drew close to my horse, and answered me in a very low voice. Now the Albanian was at least three hundred paces from us, and the Greek's conduct is only indicative of the system which prevails here. No doubt people will be cautious of using their tongues in a country where an indiscretion in speech may cause a man to lose his head. Yet after all the ráya walks

[3] The arbutus, which, like most other plants and trees, has preserved its ancient name, *κόμαρος*.

[4] Κεραμούτι. [5] Ἀμυγδαλοκεφάλι. [6] Μακεριανά.

[7] Βαβουλιανά. [8] Κουνονή.

more erect, and lives in a less horrible and wretched state of fear and trembling, under Mehmét-Alí, than in any part of what may still be called the Sultan's dominions. Here the Cretan Mohammedan fears the established authorities just as much as the Christian does: while, under the Sultan, the Musulman is an insolent tyrant, and the Greek a wretched slave.

Soon after leaving Kunoné the mountains become barren, although they still afford pasture to the numerous flocks of the district. At ten minutes before eleven a slender streamlet, shooting over a rock and falling about sixty feet, forms a picturesque waterfall[9]. We continue to ascend a little further, and then follow a nearly level path till we arrive at a little ruined chapel, where we mistake our way by taking a turn to the left. Following this wrong track all semblance of a road soon vanishes, and we have nothing left but a path, trod only by sheep and goats, and so narrow that a single false step would inevitably precipitate both horse and rider down the steep side of the mountain. We ramble along for some time without well knowing whither we are going: at length we halt, and Maniás, after using his lungs three or four times, with a force which would have rivalled even Stentor's brazen throat, is at last answered by a shepherd at a considerable distance in the mountains, and whom, for awhile, we cannot discern[10]. At length

[9] The cascade of the Styx, the most considerable in Greece, is well described by Colonel LEAKE, Travels in the Moréa, Vol. III. pp. 160—168.

[10] This τηλολαλία, among the mountains of Greece, is spoken of by Colonel LEAKE, Travels in Northern Greece, Vol. II. p. 372. and Vol. I. p. 279. "It is curious to remark with how much ease this τηλολαλία, or distant conversation, is carried on. It is an art which, as well as that of τηλοσκοπία, or of distinguishing distant objects, is possessed by the Albanians and mountaineers of Greece in a degree which seems wonderful to those who have never been required to exercise their ears, eyes, and voices to the same extent. The same qualities were among the accomplishments of the heroic ages of Greece, the manners and peculiarities of which have never been extinct in the mountainous and more independent districts of this country." ZALLONY, Voyage à Tine etc. p. 77. (quoted by SINNER, on Buondelmontii Insulae

he approaches and becomes our guide, conducting us till we come in sight of the village of Sklavopúla[11], a little before two o'clock. A steep descent and then a short ascent bring us to this village at twenty minutes past two.

The name Sklavopúla reminds me that there is a place called Sklaverokhóri in the eparkhía of Pedhiádha, and another called Sklávus in Sitía[12]. Sklavokhóri is also a name which is sometimes met with[13], and has precisely the same meaning as Dulopolis in ancient Greek. That there was a Cretan city called Dulopolis

Insulae Archipelagi, p. 185.) says, "Les Tiniens ont la voix forte et animée; et deux habitans à une distance d'une demi-lieue, même plus, peuvent *très-facilement* (!) s'entendre, et quelquefois *s'entretenir.* (!!)"

11 *Σκλαβοπούλα.* 12 Σκλαβεροχώρι and Σκλάβους.

13 See Colonel LEAKE's Travels in the Moréa, Vol. I. p. 136. where the author observes: "its name appears to indicate that it was the principal settlement of the Slavonic colonists of the *Laconice* in the middle ages." This explanation of the name is in all probability true as regards the Peloponnesian village. On the very interesting question of the influence of Slavonian conquerors and immigrants on the population of the Moréa, during the middle ages, I must again refer to a work, which I have cited in the previous volume, and which, I fear, is but little known in this country: FALLMERAYER, Geschichte der Halbinsel Morea waehrend des Mittelalters. On the influence of other northern immigrants, in Attica, see the same author's Entstehung der heutigen Griechen, published at Stuttgard and Tuebingen in 1835. In the Moréa the "Sklavokhória" may have been originally Slavonian settlements; but such an explanation can hardly be applied to Crete. Several islands near Attica, as Salamis, Aegina, Poros, Spézzia, and Hýdhra, have been colonized by Albanians, and Tzerígo has even obtained a Slavonic name: see FALLMERAYER, die Entstehung &c. p. 49. and LEAKE, Travels in Northern Greece, Vol. III. p. 70. But certainly in Crete almost all the names of villages are still Greek, and since the island possessed an ancient city Dulopolis, we are disposed to believe that it may have been the origin of the Sklavokhório or Sklavopúla of the present day, just as the ancient Port Phoron (*φώρων*) in Attica has been replaced by a Kléfto-Liméni. The words slave in English, esclave in French, schiavo in Italian, and sclave in German, were derived from these Slavonians. When they were made prisoners of war, by any neighbouring nation, they were reduced to a condition which their national name alone served to express: see ADELUNG's Woerterbuch, Vol. III. p. 1761. in the word *Sclave.* The usage of the word was soon extended from this limited sense, and came to designate any captive. MATTHEW PAR. 1252. "Cum Christianis sclavis, sic namque vocantur captivi." DU CANGE, Glossarium ad Scriptores mediae et infimae Latinitatis, Tom. VI. p. 238. ed. Paris, 1736.

we are told by several authors[14]; but unfortunately none of them in speaking of it, give any hint which might serve to determine its situation. It is said to have contained a thousand male citizens[15].

In the absence of all direct evidence on the subject, the following is a conjecture respecting the probable site of this ancient city. It is well known that the Dorian lords lived solely on the labour of their slaves or subjects, and derived all their means of subsistence from the land. Wealth could be obtained, by members of the subject classes, from commerce alone, for success in which the numerous ports of the island afforded many opportunities and advantages. What can be so likely as that some upstart town, which had thus sprung into importance, should have been called Dulopolis by the haughty and insolent Dorian freemen of the island? This conjecture seems to derive additional probability from a verse of Cratinus[16], from the name of another ancient city Poneropolis, and from the Gaidhurópolis, which we have already met with[17]. On what part of the coast Dulopolis should be placed is undoubtedly a matter which, unless some additional mention of it should be found, can never be determined with absolute certainty. I may, however, observe that there is a long tract near the shore, extending from the middle of the Sfakian district to the eastern end of Hághio Vasíli, on which I have not succeeded in establishing, on the testimony of ancient authors, a single city. That at least two or three must formerly have existed within

[14] It was spoken of by SOSICRATES, in the first book of his work on Crete: SUIDAS, Ἐστὶ καὶ ἐν Κρήτῃ Δουλόπολις, ὡς Σωσικράτης ἐν τῇ πρώτῃ τῶν Κρητικῶν. HESYCHIUS, Δούλων πόλις. Εὔπολις ἐν Μαρικᾷ, κώμην γὰρ λέγει Δούλων πόλιν. ἔστι δὲ ἐν Κρήτῃ καὶ Λιβύῃ.

[15] STEPHANUS BYZANT. Δούλων πόλις—φασὶ καὶ κατὰ Κρήτην Δουλόπολιν εἶναι χιλίανδρον.

[16] STEPHANUS BYZANT. l. c. Σημειωτέον δ' ὅτι Κρατῖνος ἐν Σεριφίοις πόλιν δούλων φησίν—

Εἶτα Σακὰς ἀφικνῇ καὶ Σιδονίους καὶ Ἐρεμβούς
ἔς τε πόλιν δούλων ἀνδρῶν νεοπλουτοπονήρων.

[17] See above, Vol. I. p. 82.

those limits is certain, and Dulopolis may have been one of them.

At Sklavopúla we obtain directions about the road, of which Maniás knows but little in this district of Sélino, invaluable as I have found his knowledge and talents as a guide in all other parts of the island. We descend gradually from Sklavopúla to Pelekánas, where we arrive at twenty minutes past three. Here we enter a valley which runs down to the African sea. Soon after four we pass a few cottages, called Tzalianá, and inhabited by Sfakians, who descend with their flocks for the winter months, and have not yet returned to Sfakiá. I learn from them that there are ancient remains on the summit of a hill to the south-west of Tzalianá, and that at Spaniáko are many tombs[18], where the Greeks excavated for ten days during the revolution, and found three gold ear-rings and several silver coins. We follow the course of the valley till twenty minutes before five, when we cross the ridge of hills on our left and descend into a similar valley, where the banks of the river are covered, as is generally the case in Crete, with lofty platanes. We follow this valley a little way, cross it, ascend the hill before us, and, sometime before sunset, come in sight of Sélino-kastéli, to which we immediately descend. I had supposed the old Venetian fort to be now occupied by Turks, and surrounded by a small village, as at Kísamo-kastéli, but found it entirely in ruins and uninhabited; there is a house, however, hard by, which is not unfrequently used as a corn magazine, and, fortunately enough for us, so employed at this time by the master of a Turkish schooner now at anchor in the bay. She arrived two days ago from Khaniá with a cargo of corn, which the proprietor has already stowed in the house, and is selling daily to the peasants of the neighbourhood. The Sfakians and Seliniotes, especially the former, thus purchase the greater part of their corn.

[18] Μνήματα.

We were glad to learn that the schooner was Turkish, since there was a chance of our finding on board her what one may call some Christian food, though no oriental Christian considers meat to be such at this season. We should have had no more chance of meeting with mutton on board any Greek vessel, during the continuance of their Lent, than we should have had of obtaining pork at any time from a Mohammedan. The Turk had bought a lamb the day before, a considerable portion of which he immediately bestowed on us; so that we fared better than, an hour before sunset, we had expected. A Sfakian shepherd whom we had passed on the road as he was tending his flock, brought us an ample supply of milk.

Sélino-kastéli, April 27.

Soon after sunrise I walked round what is called Sélino-kastéli: it consists of the remains of the old Venetian fort, which is celebrated in the history of the revolt headed by Varda Kalerghi in 1332, when it was taken by assault. It is about fifty paces square: the height of its walls, where they are now standing, is about thirty feet. There is a good engraving of the ancient fort and bay in Dapper's work[19]. I was told last night that there are remains of antiquity at Khádros, near Kántanos, or Kándanos[20], as it is pronounced, and also at a spot called Hághio Kýrko[21], near Cape Flomí. From several Sfákians, who arrive with their asses to purchase corn, I learn that there are hellenic remains near Haghía Ruméli, and that "the ancient city of the Hellenes" is above Samaría. They add: "it is said that a treasure is there concealed, but no one has yet discovered it[22]."

In Crete, as well as throughout the east, there is a universal belief that ancient sites contain hidden treasure,

[19] DAPPER, Description exacte des Isles de l'Archipel, p. 415. ed. Amst. 1702.

[20] Εἰς τὴν Κάντανο.

[21] Εἰς τὸν ἅγιον Κυριακόν I suppose must be the proper name of the place: it is always pronounced Κύρκο by the people.

[22] Ἡ παλαιὰ χώρα τῶν Ἑλλήνων εἶναι ἀπάνω ἐπὶ τὴν Σαμαρία, καὶ λέγουνι ὅπως ἔχει ἐκεῖ θήσαυρον, ὅμως δὲν εὑρήθηκε.

to find which is the object of every European's examination of them; and this idea, in some parts of the Turkish empire, opposes considerable obstacles to the traveller's researches. I had with me in Crete a pocket compass, which, on more than one occasion, a Greek noticed, and would exclaim, with an air of acuteness, and perhaps of pride on account of his discernment, "O, I know what that is for: it points out to you where the treasure is hid, when you go to any Palaeókastron."

Yesterday and to-day I saw several Sfakians, as I have mentioned. Up to this time I had scarcely spoken to any one of the inhabitants of the White Mountains, excepting my guide Maniás; and I had feared that the difference between his language and that of the other Greeks of the island, might be, in some degree, owing to an individual peculiarity: that as Alcibiades called a man a flatterer when he meant to call him a crow, so Maniás might call him a crow when he meant to call him a flatterer[23]. I found, however, that there is no difference between his dialect and that of his fellow Sfakians: the same peculiarities distinguish them all.

After leaving Sélino-kastéli at a quarter past nine, we cross a river less than half a mile to the east: the ground about it is covered with pieces of broken pottery, the only material of any kind which could possibly indicate this as the site of an ancient town.

The anonymous Coast-describer places Kalamydes between Kriu-metopon and Lissos, a city which, as we shall hereafter see, was situated a few miles to the east of Sélino-kastéli; so that this strip of land, near the ruins of the Venetian fort, might be supposed to be the site of Kalamydes, if it is an ancient site at all; but the fragments of pottery may be mere remains of the Venetian *borgo*[24], which existed outside the castle

[23] ARISTOPHANES, Wasps, 44.

Εἶτ' Ἀλκιβιάδης εἶπε πρός με τραυλίσας
ὁλᾷς; Θέωλος τὴν κεφαλὴν κόλακος ἔχει.

As the Athenian changed ρ into λ, so the Sfakian changes λ into ρ.

[24] DAPPER, Description exacte des Isles de l'Archipel, p. 415.

walls, and probably no ancient city ever existed here. At all events, if we ought not to take for granted, without a clear necessity, that Sélino-kastéli is any ancient site, still less can we admit that "Lissos, according to Ptolemy, near Criu-metopon, is Castel Sélino[25]," for we shall soon find evidence that Lissos was situated elsewhere.

About ten minutes further to the eastward, we cross a second river which issues out of another defile. The plain running down to the shore is narrow: it is bounded to the north by the mountains of Sélino, through defiles in which the streams above mentioned flow: behind Sélino-kastéli this plain may be three quarters of a mile in width: here at the second river it is not above two hundred yards. After crossing this river we have an ascent of eighteen minutes: it is so steep that we are obliged to dismount. We soon arrive at the winter-dwelling of another Sfakian mountaineer, who confirms what I have already learnt respecting the existence of ancient remains at Hághio Kyriakó, or Kýrko, and says, that the summit of a hill near us, where there is a church of the Prophet Elías, is surrounded by ancient walls: but I have made so many painful ascents, which have ended only in the discovery of petty Venetian forts, that I determine to go on towards Hághios Kýrkos, which is on the shore, and where, I think, from what is told me, that the chance of finding ancient remains is much greater than at the other place. I stay conversing with the Sfakian upwards of half an hour, and then pursue my journey: pass a sheepfold, also belonging to Sfakians, and following a road, which, like many others in this island, is so bad as to be almost impassable, arrive, at half past twelve, at a spring, and, before one o'clock, reach the small village of Prodórmi.

At Prodórmi we are welcomed by a Mohammedan, Mehmét-agá, an acquaintance of Maniás. The poorest Mohammedan of the island is called Agá by his neigh-

25 Dr Cramer, Description of Ancient Greece, Vol. III. p. 377.

bours of equal rank, and by every Christian, though of course he is not so styled by his superiors. The Mohammedan did not invite us *into* his house, but brought out stools, which he placed in the open space before it; and, in the course of a few minutes, his wife, whom we did not see, had employed herself so busily within doors, that our host put before us a dish of fried meat and eggs, a plate of olives and a bowl of milk. It may be observed that this reserve, as to Mohammedan women, is rare in Crete, and only exists with regard to strangers, and sometimes is not observed even as to them.

After our repast and half an hour's conversation with our Mohammedan host and his children, Mehmét and Ibrahím, I obtained a guide to conduct us to Hághio Kýrko, where we arrived at three o'clock. The road was terrible: we were not only obliged to descend on foot, but I almost despaired of our horses escaping without some serious accident, as they kept sliding and clambering over the rough rocks which they had to pass, when we began to descend towards the shore. On coming to the worst part of the descent we saw before us the church of Hághio Kýrko, and the site of the ancient city, in a small plain running down to the sea, surrounded on every other side by rocky hills, and presenting, on a large scale, a form very much like that of a Greek theatre.

Near the sea, where the width of the plain contracts to about one hundred and fifty paces, is a church of the Panaghía, which seems to stand on the site of an ancient temple. I observed near it fragments of the shafts of granite columns, and in the walls are several pieces of white marble, some of which seem to have been parts of pediments; others of entablatures.

The inference that this may have been a fane of Dictynna, will not fail to suggest itself to the reader.

On the south-western side of the plain, the lower part of the hills is covered with sepulchres, not scooped out of rocks, but each of them a small building, the interior of which is eight or nine feet long and six or seven wide, and the crown of its arch about six feet high. There are perhaps fifty of these tombs: their entrance-doors are so small as only just to afford room for any one to pass into them. The interior of many of the tombs is covered over with plaster.

Thus, at the site of this ancient Cretan city, scarcely a vestige of its temples can be seen, and these narrow cells are the only object which arrests the attention of the passing traveller.

> So perish monuments of mortal birth—
> Save o'er some warrior's half-forgotten grave,
> Where the gray stones and unmolested grass
> Ages, but not oblivion, feebly brave,
> While strangers only not regardless pass,
> Lingering like me, perchance, to gaze, and sigh "Alas!"

After partially examining the remains, I became so unwell that I was glad to make the best of my way to the church of Hághio Kýrko, distant about a quarter of a mile from the shore, and near which would seem to have been another temple, and a theatre.

A few observations are demanded by the peculiar situation of this ancient city, on account of the influence which such towns must have had on the Dorian institutions of the island. When those institutions flourished, "To be free from all labour save warlike exercises, to live upon the toil of his subjects and slaves[26], to know no care but the defence of his station, was the glory and happiness of the citizen; and to secure to him the enjoyment of these privileges, was the main object of all the institutions of the state[27]." It is obvious that the situation of the subject-classes, with respect to their Dorian lords, must have been different in the inland

[26] Thus at the present day the Musulman conquerors, or converts, who form the ruling class in Albania, observe, as at Kardhíki, "that the use of the musket is their only art, and their only property: the care of their fields and flocks they leave to the Christians." LEAKE, Travels in Northern Greece, Vol. I. p. 60.

[27] THIRLWALL, History of Greece, Vol. I. p. 284.

agricultural districts, and in the maritime states. The region of commercial enterprise, from which the institutions of the country excluded all the members of the territorial aristocracy, was open to their subjects; and thus probably there were never heard at the festive board of these commercial towns, the sounds of the celebrated Cretan drinking song[28]: "My great wealth is my spear, my sword, and my stout buckler, my faithful guard: with this I plough, with this I reap, with this I press the sweet juice of the vine: this is my title to be master of the mnoa. They who dare not grasp the spear, or the sword, or the faithful buckler, fall prostrate at my feet and adore me as their lord, and salute me as the great king."

If then in every maritime city there existed a tendency to oppose and destroy the Dorian institutions[29], how much greater must it have been in a place like this, the inhabitants of which had absolutely no means of existence but those which commerce supplied? Nature had given them a good port; and, therefore, although they were cooped up, when on shore, within the narrow limits of their plain, yet there was no reason why their commerce should not flourish, as that of Aegina did, before the rise of Athenian power, and as those of the mere rocks of Hýdhra, Spétsies, and Sýra, have become considerable in modern times[30].

It remains to be shewn what was the ancient city which existed on this site. In the anonymous Coast-describer's list, are mentioned, in succession and under forms slightly altered, Phoenix, Tarrha, Poekilassos, Suia, Lissos, Kalamydes and Kriu-metopon. We have already seen Kriu-metopon, and shall soon find that if Kalamydes was not at Sélino-kastéli, it was a little inland and very near it. The situation of Suia is also

[28] A scolion of HYBRIAS: see ATHENAEUS, XV. p. 695. f. EUSTATHIUS, on Odyss. VII. p. 1574. JACOBS, Anthol. Graec. Tom. VI. p. 307. ILGEN, Carm. Conviv. Graec. pp. 102—119.

[29] See HOECK's observations on this subject, Kreta, Vol. III. pp. 447—449.

[30] See LEAKE, Moréa, Vol. II. p. 432.

incontrovertibly fixed three miles to the east of these remains, which every one will therefore infer to be those of Lissos.

My conviction of this fact was greatly strengthened on ascertaining that to the north or north-east of Sélino-kastéli there is a village called Kántanos[31]. Its distance from this place is about fourteen or fifteen miles, and the Peutinger Table gives sixteen miles as that between Cantanum and Liso.

This Cretan Lissos was an episcopal see in the time of Hierocles. The order in which he mentions it, with the other bishoprics in the western part of the island, agrees very well with the supposition that its site was on this spot[32].

Of all the towns which existed on this part of the coast, Lissos alone seems to have struck coins, a fact which agrees very well with the evidence, afforded by its situation, of its having been a place of some commercial importance. Its harbour is mentioned by Scylax: and the types of its coins are either maritime, or indicative of the worship of Dictynna, as might have been expected in this part of the island.

The obverse of the first coin[33] bears the impress of the caps and stars of the Dioscuri[34], and its reverse a

[31] 'Η *Κάντανος*, pronounced Kandanos, as observed above, p. 85. The *τ* always obtains the sound of our *d* when it comes at once after a consonant and before a vowel.

[32] HIEROCLES, Synecdem, p. 650. ed. Wessel. CORNELIUS, Cret. Sacr. I. p. 235. The order is *Κίσαμος*, *Καντανία*, *"Ελυρος*, *Λίσσος*, *Φοινήκη ἤτοι Ἀραδένα*.

[33] Which is of silver, and was first published by ECKHEL, in his Numi Veteres Anecdoti, p. 152. and Tab. X. from the Imperial Cabinet at Vienna. It would seem that both Σ and P used to be doubled, either by ancient writers or by their copyists, for, on coins, we have ΛΙΣΙΩΝ, as we have ΚΝΩΣΙΩΝ, ΠΟΛΥΡΗΝΙΩΝ, &c. Eckhel, who of course inspected carefully the coin he published, mentions the other Lissos of Illyria, and adds, "Sed fabrica et typus, utrumque ex Cretae ingenio, hujus esse insulae opus, aperte loquuntur." Every one who has seen many Cretan coins must be aware of the force and conclusiveness of this argument.

[34] See the engraving of the coins, at the head of this Chapter, p. 78. Similar caps are also assigned to the Corybantes, the guardians of the infancy of Zeus:

quiver and arrow. The egg-shaped caps allude, perhaps, to the birth of the Dioscuri from the egg, which, even so late as the time of Pausanias, was still suspended in the temple of the Leucippides at Sparta[35]. The star seen above each of the caps is easily accounted for. The Dioscuri used to appear under the form of stars to sailors during a storm at sea: as soon as they were seen the clouds dispersed and the winds fell[36]. They

Zeus: see D. CELESTINO CAVEDONI, Appendice al Saggio di Osservazioni sulle Medaglie di Famiglie Romane ritrovate etc. Modena, 1831, p. 124.

[35] PAUSANIAS, III. p. 247. Ἐνταῦθα ἀπήρτηται ᾠὸν τοῦ ὀρόφου κατειλημένον ταινίαις· εἶναι δέ φασιν ᾠὸν ἐκεῖνο ὃ τεκεῖν Λήδαν ἔχει λόγος.

[36] HESYCHIUS, Διόςκουροι, ἀστέρες οἱ τοῖς ναυτιλλομένοις φαινόμενοι. HORACE, Carm. I. 12. 27.

Quorum simul alba nautis
  Stella refulsit,
Defluit saxis agitatus humor:
Concidunt venti, fugiuntque nubes.

PROPERTIUS, I. 17. 17.

Ignotis circumdata litora silvis
Cernere, et optatos quaerere Tyndaridas?

PLINY, N. H. II. XXXVII. "Exsistunt stellae et in mari terrisque. Vidi nocturnis militum vigiliis inhaerere pilis pro vallo fulgorem effigie ea. Et antennis navigantium, aliisque navium partibus, ceu vocali quodam sono insistunt, ut volucres sedem ex sede mutantes: graves cum solitariae venere—geminae autem salutares, et prosperi cursus praenunciae: quarum adventu fugari diram illam ac minacem, appellatamque Helenam, ferunt; et ob id Polluci et Castori id numen assignant, eosque in mari Deos invocant." SENECA, Natur. Quaest. I. 1. 11. "In magna tempestate apparent quasi stellae velo insidentes. Adiuvari se tunc periclitantes existimant Pollucis et Castoris numine. Causa autem melioris spei est, quod jam apparet frangi tempestatem, et desinere ventos." Compare the HOMERIC HYMN to the Dioscuri, 14—17. THEOCRITUS, Id. XXII. 6—22. VALERIUS FLACCUS, I. 569. foll. and CONRAD GESSNER, de lunariis herbis et rebus nocte lucentibus, Tigur. 1555. pp. 6—10. where other ancient authorities are quoted. These are electrical phenomena: see PRIESTLEY'S History of Electricity, pp. 394—397. ed. Lond. 1767. Dr WATSON, in the Philosophical Transactions, Vol. XLVIII. Part I. p. 210. foll. FALCONER'S Observations on the knowledge of the ancients respecting Electricity, in the Memoirs of the Literary and Philosophical Society of Manchester, Vol. III. p. 278. foll.: and SCHNEIDER'S Eclog. Physic. Anmerk und Erlaeut. p. 144. who quotes OSTERTAG. de auspiciis et acuminibus, Regens. 1779. The appearance has been frequently noticed by modern travellers and sailors. SHAW (Travels or Observations relating to several parts of Barbary and the Levant, in fol. Oxford, 1738, p. 362.) describes an ignis fatuus seen by him in the valley of Mount Ephraim, and adds that he had observed, when at sea, "in a like

disposition

also occasionally alighted on a ship in fine weather, but always in the form of stars; and their appearance was a certain augury of good fortune[37]. The attributes of the Dioscuri, namely, the name of sons of Zeus, the birth from an egg, the egg-shaped caps, the alternation of life and death, and the dominion over the winds and the waves, have been supposed to have belonged originally to the great gods of Samothrace, from whom, in process of time, they were transferred to the Laconian brothers of Helen[38]. Now the Dorians having adopted the worship of these sons of Leda, which they found established in various parts of the Peloponnese, it ought

disposition of the weather, those luminous bodies which skip about the masts and yards of ships, and are called corpusanse by the mariners." The cuerpo santo, as it is called by the Spanish sailors, seems to have been corrupted by the English into, not corpusanse, but comazant. See Captain JOHN WADDELL's account of their appearance, in the Philosophical Transactions, Vol. XLVI. p. 11. Other mariners call them the fires of Saint Elmo, of Saint Peter, or of Saint Nicolas: see HARDOUIN, on Pliny, N. H. II. XXXVII. and Dr FALCONER, l. c. On one occasion about thirty of these lights were seen on a ship at once, v. Mémoires du Conte de FORBIN, Tom. I. p. 368. (also published in the "Collection des Memoires relatifs à l'histoire de France" of Messieurs Petitot and Monnerqué, Tom. LXXIV. and LXXV. and quoted in Philosoph. Trans. Vol. XLVIII. p. 213.) LA MOTRAYE, Voyages, Tom. I. Ch. XVIII. p. 392. ed. La Haye, 1727. "Nous essuyames une terrible tempête—Un feu celeste, que les anciens appelloient Castor et Pollux, et que les Grecs appellent de Sainte Hélène, et les Latins de St Elme, se posta au haut de nôtre grand mât—Ce feu disparût bientôt, sans faire aucun mal. Un deluge d'eau, que le ciel versa ensuite, abaissa considérablement le vent." These attributes of the Dioscuri will remind the reader of the New Testament of the ship of Alexandria, which wintered in the island of Melite, and "whose sign was Castor and Pollux:" ACTS, XXVIII. 11.

37 Thus they both appeared on the ship of Lysander as he was putting out to sea on his way to engage the Athenians, on the occasion when he gained his great victory: PLUTARCH, Tom. I. p. 439. *Ἦσαν δέ τινες οἱ τοὺς Διοσκούρους ἐπὶ τῆς Λυσάνδρου νεὼς ἑκατέρωθεν, ὅτε τοῦ λιμένος ἐξέπλει πρῶτον ἐπὶ τοὺς πολεμίους, ἄστρα τοῖς οἴαξιν ἐπιλάμψαι λέγοντες·* and see HEMSTERHUSIUS, on Lucian's Dialogues of the Gods, Dial. XXVI. p. 282.

38 SEXTUS EMPIRICUS, IX. §. 37. *Καὶ τοὺς Τυνδαρίδας δέ φασι τὴν τῶν Διοσκούρων δόξαν ὑπελθεῖν πάλιν* (or rather *πάλαι*, with Hemst.) *νομιζομένων εἶναι θεῶν.* See HEMSTERHUSIUS, Lucian, Tom. I. p. 282. MUELLER's Dorians, Vol. I. p. 423. Eng. trans. But consult also LOBECK, Aglaopham. p. 1230–1.

not to surprise us to find their correligionaries, as the Cretans may well be termed, adopting it at Lissos; where therefore their emblems on coins are doubly appropriate, first, as in a maritime city[39], and, secondly, as in a state probably Dorian in its origin[40]; or at least so surrounded by Dorian colonies, and by places connected with the Dorian religion, as Elyros and Tarrha, that our meeting with the types of the Dioscuri, although found in no other Cretan city, does not excite surprise.

On the second coin the caps and stars of the Dioscuri are replaced by a dolphin, and, instead of a quiver and arrow, we see a female head, probably that of Artemis or Dictynna. Thus the maritime symbols, and those of the Cretan goddess of the chase are united as before, and shew that the coin must have belonged to this maritime Cretan city. I need not observe how the type of this goddess, found on the reverse of the second medal[41], agrees with what we have seen of her worship at Dictynnaeon, Rhokka, and elsewhere in the western part of the island.

[39] CATTANEO, Lettera al Sestini etc. p. 15.

[40] The Polyrrhenian district extended to the south sea, and Polyrrhenia was colonized by Achaeans and Laconians: (see Vol. I. p. 48. and above, pp. 47 and 49.) who are thus brought close into the neighbourhood of the Lissians. Elyros and Tarrha were also near Lissos, so that we cannot wonder that the Dioscuri should have settled here.

[41] This second coin exists at Milan, in the royal cabinet of medals at the Brera palace, where I myself examined it. The Signor Gaetano Cattaneo, Director of the Cabinet, has written an essay on this and another coin of the collection, in the shape of a letter addressed to the celebrated numismatist Sestini, now no more, intitled: "Lettera al Signor Domenico Sestini, Direttore del Museo Numismatico di S. A. I. la gran Duchessa di Toscano, sopra due medaglie greche del gabinetto reale di Milano. Di G. C. Milano, 1811." A Frenchman, Monsieur du Mersan, criticized this essay of the Signor Cattaneo, and, by doing so, drew from the Italian numismatist an answer intitled: "Difesa della Lettera di G. C. al Signor Domenico Sestini sopra due medaglie greche del Gabinetto reale di Milano contro un articolo de Signor T. Du Mersan inserito nel giornale letterario intitolato Magazin Encyclopédique del mese di Ottobre, pag. 417. Milano, Nella stamperia reale 1811." In this little work the incensed Italian shews plainly enough his French critic's ignorance of even the elements of numismatics.

If the coin belonged to the Illyrian Lissos, it would be difficult to account for the presence of Dictynna. Undoubtedly her worship spread from Crete to various other places. We find it in Rhodes, Lesbos, Thasos, Euboea, and other islands. She naturally adopted the habits of the islanders among whom she first dwelt, and became a sea-goddess, like Artemis herself[42], accompanying Apollo Delphinios[43] to preside over the altars which were dedicated to them in many maritime towns and colonies[44]. Temples were erected to her in Aegina, at Sparta, and near the Taenarian promontory[45], and she is represented on the coins of Lacedaemon[46], just as on those of Cydonia itself. She was worshipped on the Corinthian gulf[47], and, after having settled among the Phocaeans, on the coast of Asia Minor[48], she accompanied their colony to the distant shores of Gaul[49]. But, widely spread as her worship became, she was not honoured on the coast of the Adriatic, and the Dictynna

[42] HESIOD, Theog. 413.

Πόρεν δέ οἱ ἄγλαα δῶρα,
μοῖραν ἔχειν γαίης τε καὶ ἀτρυγέτοιο θαλάσσης.

EURIPIDES, Hippol. 228.

Δέσποιν' ἁλίας Ἄρτεμι Λίμνας.

[43] The notion connected with whom, from the time of the Homeric Hymnist, was invariably that of a sea god: as to what it may have been in earlier times, see MUELLER, Aeschylos Eumeniden, p. 140.

[44] PLUTARCH, de solert. animal. p. 984. a. (Tom. IV. p. 989. ed. Wytt.)

[45] PAUSANIAS, in different passages cited by HOECK, Kreta, Vol. II. p. 177.

[46] HOECK, l. c.

[47] PAUSANIAS, X. p. 890.

[48] PAUSANIAS, VII. p. 529. RAOUL-ROCHETTE, Colonies Grecques, Tom. III. p. 94. HOECK, l. c.

[49] As appears from an inscription found at Marseilles,

ΘΕΑ ΔΙΚΤΥΑ
ΔΗΜΟC ΜΑCC

See HOECK, (Kreta, Vol. II. p. 178.) who refers to the Mémoires de l'Instit. Nat. (Litter. et Beaux Arts,) An. IV. Tom. I. p. 170.

of Dalmatia never had an existence except in the imagination of a late Roman poet[50].

The argument in favour of the Illyrian city, derived from the letters seen on the dolphin of the second coin, and which are supposed[51] to denote Dalmion[52], or Dalminion[53], falls to the ground when it is remembered that the barbarian inhabitants of that country were as unacquainted with the use of money[54] as they were with the worship of Dictynna.

These are considerations against which the greater historical celebrity of the Illyrian Lissos cannot for a moment weigh[55].

On arriving at the church of Hághios Kyriakós, which the people called Kyrkos, we found, in a shed near it, the Papás of a neighbouring village, who, possessing some of the land of this little plain, was come to stay here a few days. He had brought with him no provisions, except some barley bread of the worst kind and a little indifferent oil, which latter article is to himself a forbidden luxury at present, it being now *the great week*[56] of the Greeks. For supper, then, a few herbs were boiled by the priest, and were eaten by

[50] CLAUDIAN, de Sec. Cons. Stilich. v. 302.

> Dalmatiae lucos, abruptaque brachia Pindi,
> Sparsa comam Britomartis agit.

[51] By SESTINI.

[52] STEPHANUS BYZAN. v. Δάλμιον. EUSTATHIUS, on Dionys. Perieg. v. 95.

[53] STRABO, VII. p. 315. Δαλμίνιον δὲ μεγάλη πόλις ἧς ἐπώνυμον τὸ ἔθνος. EUSTATH. l. c.

[54] STRABO, X. p. 315. Ἴδιον δὲ τῶν Δαλματέων, τὸ διὰ ὀκταετηρίδος χώρας ἀναδασμὸν ποιεῖσθαι· τὸ δὲ μὴ χρῆσθαι νομίσμασι πρὸς μὲν τοὺς ἐν τῇ παραλίᾳ ταύτῃ ἴδιον· πρὸς ἄλλους δὲ τῶν βαρβάρων πολλοὺς κοινόν. See also the SCHOLIAST ON DIONYSIUS PERIEG. v. 97.

[55] SESTINI lets it weigh against some of them in his Descrizione d'alcune Medaglie greche del museo del Signore Carlo D'Ottavio Fontana di Trieste, per Domenico Sestini, p. 35. where he concludes, "Dai geografi poi si parla d'un Lissus dell' Illirico più celebre del Cretico, e non saprei decidermi, se per la nostra medaglia debba preferire quello o il Cretico."

[56] Ἡ μεγάλη ἑβδομάδα.

himself, Maniás, and my companion: for my own part I lay down supperless[57].

[57] MATRON, in Athenaeus, IV. p. 134. f.

"Ενθ' ἄλλοι πάντες λαχάνοις ἐπὶ χεῖρας ἴαλλον,
ἀλλ' ἐγὼ οὐ πιθόμην!

Would that I had refrained in order to enjoy such a feast as followed the herbs in the house of Xenocles! We should have been glad indeed to have had, now, even that lenten fare of salt fish and sea-urchins, which Matron there disdained to touch:

'Ωμοτάριχον ἐῶν χαίρειν Φοινίκεον ὄψον·
αὐτὰρ ἐχίνους ῥῖψα καρηκομόωντας ἀκάνθαις.

RHODHOVANI AND THE SITE OF ELYROS.

CHAPTER XXVII.

WALK TO SUIA. VULTURES, EAGLES, FALCONS, AND THE COCK OF THE WOODS IN CRETE. KRUSTOGHERAKO. AFRICAN HOUSEHOLD SLAVES. VILLAGE OF RHODHOVANI. SITE AND REMAINS OF ELYROS. THALETAS. HONEY OF RHODHOVANI. COINS OF ELYROS. RIDE TO TEMENIA.

April 28.

After a night spent without obtaining much sleep, I found, at day-break, that I was covered with traces of the ravages committed on me by the light troops, which, in this part of the world, invariably take up their quarters in unfrequented buildings, even more than in inhabited ones; and whose attacks had been kept up, with the most annoying perseverance, throughout the night. I was still so weak and unwell that I hardly knew what course to adopt, when I learnt that the direct road to Súia, about three miles further to the east, and also on the shore, is absolutely impassable for horses. The whole distance by the paths, along which the road is best, and which may be traversed on horseback, was said to be about twelve miles. My only alternative, therefore, was between walking on foot for an hour, directly over the hills, and making, on horseback, a tedious circuit, part of which would be along our yesterday's road. I determined on walking, and on sending the horses round. The ascent up the hills does not take above twenty minutes, but the road is such that I was glad not to have attempted bringing our steeds along with us, although asses, which are never shod in this part of the world, do constantly traverse it without any danger. When we reach the top of these hills bounding the little plain of Lissos on the east, summits of mountains still covered with snow are visible four or five miles in an east and north-east direction. After starting a covey of partridges, we disturbed a flock of ten large vultures, of a light brown colour, with wings which were nearly black, as they were feeding on the body of a kid lying near our path.

Belon observes that vultures, eagles and falcons, in Crete, build their nests, not, like other birds, in trees, but in difficult and precipitous places, among rocks which look towards the sea[1]. He mentions the fact, of which we have here ocular evidence, that vultures

[1] BELON, Singularitez etc. Liv. I. Ch. XI. fol. 13. His phrase is: "en lieu moult difficile et precipiteux." He subjoins: "Les voulants desnicher, faut auoir vne longue corde qu'on laisse pendre le long du roc, dont le bout est attaché dessus le faiste de la montagne à quelque pau fiché en terre. Un paysan deualle le long de la corde jusques à tant qu'il paruient au lieu où est le nid du Vautour, puis retourne à mont par la mesme corde par laquelle il estoit descendu. Autrement mettent vn petit garson dedans vne grande corbeille, qu'ils deuallent le dessus le roc contre bas: & quand il est parvenu au nid, lors il met les oiseaux en sa corbeille & se fait retirer à mont." The eagle's building in high places is frequently alluded to in the Old Testament, as in JOB, XXXIX. 27–28. Our MILTON says, (Paradise Lost, VII. 423.)

There the eagle and the stork
On cliffs and cedar-tops their eyries build:

I suppose him to mean cliffs for the eagle, and cedar-tops for the stork. See PSALM CIV. 16–17, and BOCHART'S Hierozoicon, Vol. III. pp. 89–90.

frequent the mountainous parts of the island, and adds, that they destroy young lambs, kids and hares. In his time the vulture's wings used to be sold to the arrow-makers, and its skin to the furriers[2].

It is singular that the well-known cock of the woods of northern Europe[3], distinguished by the name of capercailzie, in Scotland, where the species has become extinct only within the last fifty years[4], should have been seen in so low a latitude as Crete. The fact, however, is recorded by Belon, in a manner which seems to leave no doubt on the subject[5].

After leaving the vultures, we descend and soon arrive at Súia, which, like Hághio Kýrkos, is entirely uninhabited. Some peasants shewed us a spring of brackish water, and conducted me to a mutilated inscription[6], from which alone it would have been impossible to determine the name of the city[7].

The modern name Súia, or, if I adopt an English instead of the Italian mode of writing the word, Soúia, is the same as that of the ancient city[8], of which we

[2] BELON, l. c. "Ils les escorchent, & en vendent les œlles aux artilliers, qui s'en servent à faire des ampennons aux flesches: & la peau aux pelletiers qui la conroyent pour en faire fourrures, qui sont vendues bien cher."

[3] I am told, by an excellent authority on the subject, Lord Melgund, that this noble bird is frequently seen by travellers in Norway, still oftener in Sweden, and is also common in the mountains of Bohemia. It finds its way, every winter, to the tables of the wealthy in London, and is frequently met with in the bill of fare of the Parisian restaurant.

[4] Attempts have been lately made to re-naturalize the bird in Scotland.

[5] BELON, Singularitez etc. Liv. I. c. IX. fol. 11. "L'oiseau que les Romains nommèrent tetrao, et lequel les Italiens nomment pour le iour d'huy Gallo Cedrone, et en Auvergne un faisan bruyant, et en Savoye un coc de bois, *est souuent veu* par les forests des hautes montagnes de Crete, deux fois plus gros qu'un chappon."

[6]
ʹΡΡΑΝΙΟΝΙΟΥ
ΥΝΔΟΥΥΙΟΝ
ΞΙΩΝΗΠΟΛΙΣ

[7] We can obtain a termination in ΙΩΝ neither from Συϊάτης nor Συϊεύς, the two forms, given by Stephanus of Byzantium, for the name of the citizens.

[8] The present name of the place in modern Greek is Σούϊα; that of the ancient city was Συΐα. Although the number of words in which the ῡ has been

learn from Stephanus of Byzantium, in a passage which I have copied into a note-book, and have with me when I find these remains, that it was the port of Elyros[9]. Now Pausanias says expressly that Elyros existed in the mountains of Crete even in his days, and therefore I feel satisfied that I ought to find vestiges of that most interesting ancient city somewhere in these mountains of Sélino, a few miles from the sea. The anonymous Coast-describer speaks of a city Syba or Syva[10], evidently meaning to indicate the same Syia as being sixty stades to the west of Poekilassos, which is therefore situated to the east, somewhere within the confines of Sfakiá. The common boundary of Sélino and Sfakiá is said to be at the church of the prophet Elías and

been thus altered into ου̅ in the modern Greek is not very great, yet they will suffice to shew that the change does not unfrequently happen. The name of *Σκαρδαμούλα* thus fixes the site of *Καρδαμύλη* in Laconia, as I have already mentioned. I have also been told, in England, that the island *Γύαρος* is commonly called *Γούαρος*, another instance of the same law. In the modern language *χρυσὸς* and *χρουσὸς* are both in common use: *πτερούγα* also is met with. A song which I learnt at Hierápetra begins:

Κάτω εἰς τὴν κάτω γειτονιάν, πέρα εἰς τὴν πέρα ῥούγα,
χαμαιπετᾷ μιά πέρδικα μὲ τὴν χρουσὴν πτερούγα.

We have similarly, in the Erotocritos, the verse

Τὰ φρούδια τζὴ ἐνεσίκωνε μὲ μαστοριάν ἡ κόρη.

From *Ἄχυρα* it appears that a diminutive *Ἀχούρια* (instead of *Ἀχύρια*) has been formed, and is used as the name of villages in the Peloponnese. LEAKE's Moréa, I. p. 90. II. p. 252. This corruption existed in the word *σοῦκον*, for *σῦκον*, at Athens, in the time of Theodosius Zygomalas: (see CRUSIUS, p. 99. DU FRESNE, 1410.) and it prevails more in some parts of Greece than in others at the present day. A Cretan chieftain was once talking to me about the celebrated Rumeliote leader Khadjí Mikháli, and of the distrust and misunderstanding between him and the Sfakians. In fact, said my informant, the Khadjí had lived so much among the Turks in Rúmeli, "that he did not know how to speak Greek, and used to say *τρούπα* for *τρύπα*." Thus it is stated that *Κύρα* is corrupted into *Κούρα* at Naxia: HAWKINS, in Walpole's Memoirs, Vol. II. p. 395: and, generally, the ου̅ is "a favourite" in the language, as Colonel LEAKE, Researches, p. 71, has observed. This ου̅ has even usurped the places of *ευ* (see KORAY, ATAKTA, Tom. II. p. 430.) and of οι̅, as in *Προικίον*· (see KORAY, ATAKTA, Vol. IV. p. 273.)

[9] STEPHANUS BYZAN. v. *Συΐα*.

[10] AUCTOR STADIASMI, in Geograph. Graec. Minor. Tom. II. p. 496. ed. Gail.

of Hághios Isódhoros, about three miles to the east of Súia.

I learn that, till a few years ago, there was here a large marble slab covered with an inscription. A kaík came and took it away. In the middle of the plain is a winter stream[11], between which and the rocky cliffs, forming the eastern boundary of the plain, are found the existing remains of the ancient city. I see spread over the ground here pieces of pottery and three fragments of the shafts of columns. Although the plain is narrow even at the sea, and contracts at no great distance from it, yet Súia is not so cooped up by surrounding rocks as Hághio Ký́rko. I observe remains of the city walls, and also, in several places, those of public buildings; but nothing deserves a particular description. None of the visible remains can be more ancient than the time of the Roman emperors. They are either of different sized stones or of brick work. The sculptured cross on some fragments of white marble would seem to indicate the existence of the city in Christian times, and from the mention of Elyros in Hierocles's list of Cretan cities, then reduced in number to one-and-twenty[12], there can be but little doubt that its small port also continued to exist at that time. On the side of the hill, to the south-east of the city, are several tombs resembling those of Hághio Ký́rko.

My guide does not arrive, with the horses, at Súia till near one o'clock, and has been at least four hours in performing the journey, which is less than three miles along the path followed by us, and is probably not two by sea. We reached Livadhá, a village two miles to north-north-east of Súia, about two o'clock. We found the honey at this place most delicious. After a short stay we continued to ascend towards the village of Krustoghérako, distant not a mile and a half from Livadhá, although, owing to the steepness of the ascent,

[11] Ξηροπόταμο.

[12] HIEROCLES, Synecdemus, p. 650. ed. Wessel.

we were near forty minutes in reaching it. I found an ancient sepulchral inscription in the village[13].

The villagers, in talking over the events of the war, mention as a remarkable peculiarity in the recent history of Krustoghérako, which was burnt by the troops of Khusseín and Mustafá beys, when they overran the district of Sélino, that none of its inhabitants were enslaved during the whole war.

An ancient ring of twisted gold, and a stone on which was cut a Demeter with her horn of plenty, as well as some coins, were shewn me to-day; but the price demanded for each of them was so extravagant that I could not purchase any thing. It is a common belief that the possession of an ancient coin is a sovereign charm against maladies of the eyes[14].

At last I learn, after making an infinity of vain inquiries during the whole day, in every place to which

13 ΚΑΛΛΙΠΟΝΟΣ
ΚΑΙΑΙΤΟ ΣΑ
ΑΓΗΣΙΦΩΤΩ
ΥΙΩ

14 Gold and silver medals of Alexander were supposed, in heathen times, to be of great use as a charm to ensure the prosperity of their possessor. TREBELLIUS POLLIO, Life of Quietus, (c. XIV.) "Dicuntur juvari in omni actu suo, qui Alexandrum expressum vel auro gestitant, vel argento." ST JOHN CHRYSOSTOM, Homil. XXV. ad Pop. Ant. bitterly inveighs against the practice: see GALEOTTI, Museum Odescalchum, Tom. I. Tab. XV. p. 20. The Italian Girolamo Dandini, who visited the principal city of Crete in 1599, on his way to Mount Libanus, after mentioning several remarkable things, adds, "but that which is wonderfully surprizing, and beyond the force of nature, are certain pieces of money, which they call Saint *Helen's*, that are found up and down the fields, where there is also brass and other silver. They pretend that that Saint happening to be in that country without money, made some of brass, which in passing of them changed into silver: this money, they say, hath the vertue to this day, to cure the falling sickness in them that hold it to their hand." See p. 12. of "A Voyage to Mount Libanus: also a description of Candia, Nicosia, Tripoly, Alexandretta, &c. written originally in Italian, by the R. F. Jerome Dandini. London, 1698." The chapter on Candia (pp. 9—13.) is also printed in Pinkerton's Collection. The work was translated into French, under the title of "Voyage du Liban, &c. à Paris, 1675."

I went, that there exists a palaeókastron near the village of Rhodhováni, situated some miles to the westward of Krustoghérako; and, consequently, I determine to go thither to-morrow morning with great hopes of finding the site of Elyros.

April 29.

We leave Krustoghérako at twenty minutes to seven: repass Livadhá, and continue to descend till twenty-five minutes past seven, when we cross the bed of the river, and, in another quarter of an hour, arrive at the village of Moné, where a Mohammedan, an acquaintance of Maniás, presses us warmly to stay and breakfast. We sit down with him for half an hour under the lofty trees which overshade his dwelling, and are attended by two negro slaves, who have as yet learnt but little of the language of the island.

There are but few negros in the villages of Crete, although in the principal towns there are slaves in the families of almost every Mohammedan gentleman. The price of labour is every-where very high, the difficulty of obtaining labourers in many cases amounting to an absolute impossibility, and the markets of Khaniá and Megálo-Kástron are as regularly furnished with human flesh as they are with bullocks, the supply of both being chiefly drawn from the same place, Bengázi. One may therefore wonder, that of the small proprietors, who form the rural population of the island, so few should have slaves to assist them in the cultivation of their lands.

We leave Moné at a quarter past eight, and in about half an hour see, somewhat to our right, ruins, which prove to be situated on the site of an ancient city. After a slight survey of these remains we proceed to the village of Rhodhováni, about half a mile farther on. The chief fountain in this village is a very copious source, surrounded as usual by lofty trees, and occupied, at the moment of our arrival, by seven or eight Greek women, employed in washing their linen.

In the masonry of this fountain I observe, built up along with less precious materials, three fragments of ancient sculpture, one of which, a female figure that has lost the feet and head, is well executed: the drapery is very graceful.

On sitting down under the shade of the trees by this fountain we are soon surrounded by all the villagers who were either at home or near enough to hear of the arrival of two Europeans at Rhodhováni, an event which if not sufficiently contrary to the established course of nature to be considered as miraculous, is at least unprecedented, and is regarded as most wonderful here and everywhere else in this eparkhía of Sélino.

The Hellenic city goes by the name of Kephalés[15]: the grammatikós of Rhodhováni proves an unusually intelligent and very obliging person: I gladly accept his offer of accompanying us with another villager, to point out all the ancient remains: among which I first noticed a building, consisting of a series of arches; and next, vestiges of walls, especially on the north and north-eastern sides of the ancient city. On part of the site of what seems undoubtedly to have been a Christian church of some antiquity, there exists a modern Greek chapel. The length of the old building is seventy-three feet, and its width forty-eight feet: its altar, like that of the diminutive chapel which stands on part of its site, seems to have been, as usual, at its eastern end.

At some little distance from the church and chapel are remains indicating far greater splendour, of a very remote period: a few massive stones, some pieces of an entablature, and several fragments of the shafts of columns, are all that now remain of an ancient temple. I learn from my very civil and communicative guide that, upwards of twenty years ago, there were standing "two great stones," which formed an entrance to the temple. Each of them was ten or eleven feet high, and

[15] Τζὴ Κεφαλαῖς.

several feet more were buried in the ground. Hence the spot was called Orthés Pétres[16]. They were destroyed by a Mohammedan, who converted them into millstones[17].

My informant remembers when the ground all about here was covered with mosaic, like some remains still existing in the church, formed of red, black and white marble. A further examination shews us some cisterns. On the north and north-east sides of the ruins are remains of the city walls, the circuit of which must have extended about two miles.

On the very summit, the elevation of which, however, is so slight, above the rest of the site, that one hardly knows how to call it the acropolis, are remains of walls which seem to have belonged to a fortress. The view hence is very extensive, and its beauty is such that the place well merits the epithet bestowed upon it by my guide[18]. To the south we have the valley and little plain of Súia, bounded by the African sea: to the west, a range of lofty mountains: to the east, still loftier, many of them, even now, that the season is so far advanced, being still covered with snow. The ranges of undulating hills to the north have several villages, surrounded by olives, almonds and other trees, on their lower slopes.

My excellent guide, the grammatikós of Rhodhováni, also tells me, and another villager mentions the same fact, that between this place and Súia, remains of an ancient aqueduct[19] are to be seen: it had its source to the north-west, about two miles hence, on a somewhat higher range of hills, near the village of Livádha. I learn too that there are very considerable Hellenic remains at a place called Kastrí, due west, about four miles off, and on a hill still higher than this where we

[16] To use the Cretan expression, *τζὴ ὀρθαὶς πέτραις*.

[17] The grammatikós observed, *τὶ ὄφελος ἦτον θὰ τὰ κόψωσι;*

[18] One of his expressions was *εὔμορφος τόπος*!

[19] Which he called *αὐλάκι*, and compared to that by which the city of Khaniá is now-a-days supplied with water.

now are. The villagers tell me that Spaniáko once belonged to Spain, and obtained its name from its Spanish occupants, whose nation, at that time, possessed Sélinokastéli and all the surrounding district[20].

Every scholar will have seen long ago, that all the remains which we have just examined belong to the most important ancient city Elyros. Pausanias[21] says that the city Elyros still existed in his time, in the mountains of Crete, and we also learn from him, that the Elyrians sent to Delphi a brazen goat, which he saw there, and which was represented in the act of giving suck to the infants Phylacis and Phylander. The Elyrians affirmed that these were children of Apollo and the nymph Acacallis, whose favoured lover the youthful god had been in the house of Carmanor at Tarrha. Elyros is also spoken of by Scylax[22], who places it between Cydonia and Lissos: a straight line drawn from Khaniá to Hághios Kýrkos, the sites of those two cities, would, I believe, pass through this Palaeókastron, a fact which shews the accuracy of the old geographer's description. Stephanus of Byzantium, as we have seen, mentions this city when he speaks of Súia[23], and also notices it elsewhere[24].

[20] On the presence of the Catalans in Greece, Pachymeres, Nicephorus Gregoras, Ducange, and GIBBON, c. LXII. may be consulted. Similar traditions of the presence of Spanish occupants are frequently found in Northern Greece, see LEAKE, Travels in Northern Greece. Vol. I. pp. 91. 336. and 392. Vol. II. p. 15. and Vol. III. p. 29. On the occupation of Attica by the Catalans, see LEAKE, Topography of Athens, p. LXXV. foll.

[21] PAUSANIAS, X. p. 835. Ἔστι δὲ ἐν τοῖς Κρητικοῖς ὄρεσι καὶ κατ' ἐμὲ ἔτι Ἔλυρος πόλις· οὗτοι οὖν αἶγα χαλκῆν ἀπέστειλαν ἐς Δελφούς· δίδωσι δὲ νηπίοις ἡ αἶξ Φυλακίδι καὶ Φυλάνδρῳ γάλα. παῖδας δὲ αὐτοὺς οἱ Ἐλύριοί φασιν Ἀπόλλωνός τε εἶναι καὶ Ἀκακαλλίδος νύμφης· συγγενέσθαι δὲ τῇ Ἀκακαλλίδι Ἀπόλλωνα ἐν πόλει Τάῤῥᾳ, καὶ οἴκῳ Καρμάνορος. A learned friend suggests to me the correction *κατ' οἶκον* Καρμάνορος.

[22] SCYLAX, in Geograph. Graec. Min. Tom. I. p. 265. ed. Gail. Κυδωνία, καὶ λιμὴν κλειστὸς πρὸς βορέαν· ἐν μεσογείᾳ δὲ Ἔλυρος πόλις. πρὸς νότον δὲ Λίσσα, (vulg. νότον Μέλισσα) πόλις καὶ λιμήν, παρὰ Κριοῦ μέτωπον.

[23] STEPHANUS BYZANT. v. Συΐα, πόλις μικρὰ Κρήτης, ἐπίνειον οὖσα τῆς Ἐλύρου.

[24] STEPHANUS BYZANT. Ἔλυρος, πόλις Κρήτης, ὡς Ξενίων ἐν Κρήτῃ. ὁ πολίτης Ἐλύριος.

It seems from a passage of Suidas[25], that Thaletas, the Cretan poet, lived before the time of Homer, and was born at Elyros. Cnossos[26] and Gortyna[27] are, it is true, likewise mentioned as his birth-place, but, as has been well observed[28], the popular opinion in cases of such uncertainty, chooses the more celebrated town; and the connexion of Elyros with Tarrha, and with Apollo's worship, will make us decide in favour of its claim to be regarded as this ancient poet's birth-place. It appears from the authorities quoted by Kuster on the passage of Suidas, by Tzschucke on Strabo[29], and by Meursius in his Crete, that the same poet is spoken of both as Thaletas and Thales. Strabo tells us that the Cretans used to refer to this Thaletas, who was considered as the inventor of the *cretic* rhythm, the paeans and the songs of their country, and many of their institutions[30].

After finishing our examination of the ancient site, we returned to the village of Rhodhováni. Some delicious honey, given to us by the grammatikós, reminded me of the bee seen on the old Elyrian coins[31]. My companion makes a sketch of the village and ancient site, which is on the hill to the right a little further off. The sea just beyond Súia is also visible. The snow-

[25] SUIDAS, *Θαλήτας, Κρῆς, Ἐλύριος* (adopting Meursius's certain and received emendation for the corrupt *ἢ Ἰλλύριος*) *λυρικὸς γεγονὼς πρὸ Ὁμήρου. Μέλη.* On other dates at which Thaletas is placed, see ULRICI, Geschichte der Hellen. Dichtkunst, Vol. II. p. 213.

[26] SUIDAS, *Θαλήτας, Κνώσσιος, ῥαψῳδός· ποιήματά τινα μυθικά.*

[27] PLUTARCH, on Music, Tom. V. p. 637. ed. Wytt. Oxon.

[28] By Professor HOECK, Kreta, I. 27.

[29] STRABO, X. p. 480. Tom. IV. p. 279. ed. Tzsch.

[30] STRABO, X. p. 480. fin. *Ὡς δ' αὕτως καὶ τοῖς ῥυθμοῖς Κρητικοῖς χρῆσθαι κατὰ τὰς ᾠδὰς συντονωτάτοις οὖσιν, οὓς Θάλητα ἀνευρεῖν· ᾧ καὶ τοὺς παιᾶνας, καὶ τὰς ἄλλας τὰς ἐπιχωρίους ᾠδὰς ἀνατιθέασι, καὶ πολλὰ τῶν νομίμων.* Whoever wishes for more information respecting Thaletas, as one of the ancient lyric poets, may consult the authorities indicated by ULRICI, l. c.

[31] On these coins consult PELLERIN, Recueil de Médailles etc. Tom. III. p. 63. Pl. XCVIII. fig. 20. ECKHEL, Numi Veteres Anecdoti, pp. 148, and 158. SESTINI, Museo Hedervariano, Parte Europea, p. 154. MIONNET, Supplément, Tom. IV. p. 319.

covered mountain is, I should suppose, the most easterly of the chain of the White Mountains. It is in the eparkhía of Sélino, and is the only snow-clad mountain out of Sfakiá.

I wondered at finding so few remains of the ancient city in the modern village. In addition to the sculptured fragments at the fountain, I did at last succeed in discovering an inscription: its letters, however, were worn almost entirely away, and the slab was placed upside down in the outer wall of a house, at a considerable height from the ground, so that it was very difficult to make any thing out[32].

The intelligent grammatikós is the Kiaías of the eparkhía, and is therefore the best authority on the subject of its produce. The mean amount of the seventh of oil in Sélino is three thousand místata: and all the other produce barely equals that of oil. Sélino is the poorest district in Crete next to Sfakiá.

We leave Rhodhováni at ten minutes past five, and soon cross the head of the valley which lies to the west of the village, and, ascending for about twenty minutes, pass the hamlet of Mázo. Continuing to ascend for nearly all the rest of the way, we arrive at the village of Teménia soon after six o'clock, and take up our quarters with a relation of the grammatikós of Rhodhováni: he is, however, as stupid and reserved a fellow as the Rhodhovanian was intelligent and communicative.

During the evening our host and his wife were visited by an elderly dame, accompanied by her young and

[32] The following letters were transcribed by me:

ΑΓΑΘΑΙΤΥΧΑΙ
ΚοΣΜΩΤ.ΙΕΛ
ΤΩΑΓΗΣΙΦΩΩΕΛο·ΕΤοΙ·Κο
Σ·ΙΙΛΠοΛ··ΔΗΣΤ
ΦΩΝΩΝΛΣΑΝΛ
ΙΙΙΛΙΩΝ

beautiful daughter, whose dress, peculiar to one of the islands of the Archipelago, struck me, on account of the great difference between it and that of every other woman in Crete. I learnt that she is a native of this village, but having fled to the islands[33], with thousands of her sex, for protection and security, during the continuance of the war in Crete, she had not yet abandoned, since her return, the costume of her place of refuge.

[33] Eἰς τὰ νησιά.

ANCIENT WALLS OF HYRTAKINA.

CHAPTER XXVIII.

THE SITE AND EXISTING RUINS OF HYRTAKINA.

April 30.

THE morning is beautiful as usual: the weather has long been quite invariable. We leave the village of Teménia at half past five, and are nearly half an hour in reaching the summit of the hill to the south, on the top of which I yesterday learnt that there are Hellenic remains: I find all my hopes fully realized.

To the south of this ancient site the extremity is narrow and precipitous: the ascent on the eastern side is also so steep that one looks not to find walls there. Along the south-western and western sides, however, considerable remains of the ancient walls still exist: their height above the ground varies from two to five or six feet. After following their course on the south-western side for about two hundred paces, we arrive at an entrance not only defended by the ordinary projections of the outer wall, but by what is here observed within, two other walls extending for some distance nearly parallel to the outer one. The distance between the outer and the second wall is about ten paces; that between the second and third near forty. Soon after passing this entrance, we find a considerable piece of the outer wall, which is still from five to twelve or thirteen feet high. The piece sketched[1] is perhaps rather more regular, in the size and forms of the stones, than the greater part of the walls which remain; but the whole may be considered as a near approach to what has been termed the second style of cyclopian masonry. As I pass along, following constantly the course of the walls on the north-western side of the city, I observe on the ground numerous pieces of pottery, and also notice that the stones are, in some places, more massive than those sketched. On this north-western side of the city there appears to have been a defended entrance in one of the interior walls, and between this entrance and the acropolis, of which I will soon speak, three other walls can be distinguished, although the ground is a continual ascent and is very rocky. The whole length of the present remains of walls on the south-west, west and north-west sides of this ancient city scarcely exceeds half a mile.

The little acropolis is situated on a mount about one hundred and fifty paces from the southern extremity of the site: round its base are seen remains of walls: I observe also some slight foundations on its summit. Between this and the southern extremity the ground is covered with fragments of pottery, and other foundations, probably of the walls of houses, are also

[1] The length of which is 11 feet 6 inches, its height being 8 feet 6 inches.

seen. A little to the south-west of the acropolis are remains of an entrance, one of the stone pillars of which is still standing, and a small piece of wall, consisting, like the rest, of massive stones.

From the summit of the acropolis we have an extensive view in every direction: the extremity of the promontory of Kástel-Sélino and the sea, even over the hills which terminate in Cape Krío are visible: we have before and below us the villages of Plataniás and Prodórmi: to the west and north-west are seen two other villages, half-buried in the olive-trees by which they are surrounded. On descending I notice two tombs cut out of the rock.

I believe this to be the site of the ancient city of Hyrtakos, or Hyrtakina, the coins of which, presenting as they do types similar to those of Elýros[2], might alone lead one to suppose, if we knew nothing of its situation, that it was somewhere in these mountains. But, little as we learn of its position from Ptolemy[3] and Stephanus of Byzantium[4], yet we may safely infer, from the former's words, that it was situated to the south-east of Polyrrhenia, and to the west of Lappa. Scylax teaches us something more respecting its site: he places it on the south side of the island, and to the south of the Dictynnaean temple of Artemis and the Pergamian district[5].

[2] On the coins of Hyrtakina are found the epigraphs ΑΤϘΥ, and ΥΡΤΑΚΙΝΙΩΝ. They have been erroneously ascribed to Elyros, by COMBE, (who read ΥΡΙΑΚΙΝΙΩΝ,) Numi Musei Britannici, Tab. 25. fig. 20. RASCHE, Lexicon Rei Numariae, Tom. II. Part. I. 600. and MIONNET, Description de Médailles etc. Tom. II. p. 277. Nos. 157 and 158. See SESTINI, Lettere e Dissertazioni Numismatiche, Tom. VIII. pp. 4 and 5. MIONNET, Supplément, Tom. IV. p. 324.

[3] PTOLEMY places it among the inland cities of Crete.

[4] STEPHANUS BYZANT. v. Ὑρτακός, ἢ καὶ Ὑρτακῖνος, πόλις Κρήτης. ὁ πολίτης Ὑρτακῖνος. Πολύβιος δὲ τὸ θηλυκὸν Ὑρτακίνη ἀπὸ τοῦ Ὑρτακῖνος ἐθνικοῦ. ἀπὸ Ὑρτάκου δὲ Ὑρτάκιος.

[5] SCYLAX, p. 18. ed. Huds. (compare MEURSIUS, Creta, p. 40.) Δικτυνναῖον Ἀρτέμιδος ἱερὸν πρὸς βορέαν ἄνεμον τῆς χώρας Περγαμίας. πρὸς νότον δὲ Ὑρτακίνα. Κυδωνία, καὶ λιμὴν κλειστός, πρὸς βορέαν. ἐν μεσογείᾳ δὲ Ἔλυρος πόλις.

This agrees well with the situation of these ruins; and, as he goes on to the eastward in his description, he next speaks of Cydonia and Elyros, which shews clearly that Hyrtakina was to the westward of Elyros, and thus confirms the supposition that this is its site.

WALLS NEAR KHADHROS.

CHAPTER XXIX.

THE SITE OF KANTANOS, NEAR KHADHROS. VILLAGE OF SPANIAKO, AND RUINS OF TWO TOWERS NEAR IT.

April 30, *continued.*

We return to the village of Teménia at ten minutes past nine, and then pursue the course of the river westward: the plane-trees with which its banks are covered are just shooting out their foliage. We soon leave the river on our left, and, after ascending for about a mile, descend, during nearly a quarter of an hour, to the river and village of Stráti. Here we remain till half past ten, my horse having run away as we descended the hill on foot. The banks of this river of Stráti also are covered with fine plane-trees. Immediately after leaving the village, the promontory of Sélino-kastéli, four or five miles off to the south-west, becomes visible, and we also see the course of the valley all the way down to it. In less than an hour we arrive at a shepherd's cottage to the north of Khádhros, and find that the ancient remains, of which I am in search, are situated on a conical hill about a mile to the south, and which we passed on our left as we descended into the valley of Kántanos. The site is of course on the eastern side of this valley, and is easily distinguished by the ruins of the little chapel of Haghía Iréne, which crown its summit. After obtaining a crust of bread and some misíthra at the cottage, I proceed to examine the ancient remains.

From the summit we have a fine view of the valley of Kántanos: we are three or four miles from the sea, to which the valley leads in a southerly direction: the little villages designated by the common term Kántanos are near its north-eastern extremity: the hills on both sides of the valley are well covered with olive-trees for a considerable way up their sides. Near the chapel of Haghía Iréne I searched in vain for traces of any ancient building. The hill has two peaks, and the chapel is situated on the southernmost of them: these peaks are formed by a cleft in the rock ten feet wide, forty to sixty feet high, and about two hundred paces long, in a direction north-west and south-east. One sees plainly that it is the effect of an earthquake: the irregularities of the surfaces on the two sides correspond so faithfully, that if they could again be brought together they would fit accurately: the rock is limestone, and small stalactites are formed on some of its overhanging portions.

Remains of walls, which are here, as at so many other sites, almost the only evidence now left us of the existence of an ancient city, are found on the eastern and south-eastern sides of the hill of Haghía Iréne. They are traceable for little more than one hundred and fifty paces. The style of their masonry however attests a high antiquity: and thus, in one day, I have had the good fortune to find the sites of two as yet unvisited ante-Homeric cities. I may add that, about twelve paces from these remains, are some slight ruins of an outer wall. Other portions of the continuation of both these walls are discernible to the east of the hill on the northern side of the cleft, and are perhaps even better preserved than those which I have described, and of which a specimen is engraved at the head of this chapter.

To the east of these remains, which, in the whole, run three or four hundred yards on the side of this hill, the slope is more gradual than on the other sides; and here also are discernible vestiges of four other walls, meant no doubt to strengthen the side where the city was, from the nature of the ground, most exposed to attack. I also notice several tombs, cut out of the solid rock, on the side of the hill to the east of the ancient city, and on that side below some of the tombs are also remains of a wall. Khádhros, the nearest village to Haghía Iréne, is about half a mile to the north.

We shall not find it difficult to fix, with certainty, on the ancient city which existed here. There is a district called Kántanos by the Cretans, just as we have districts of Mesóghia, Enneá-khoriá, and so forth, the appellation being used to denote collectively a number of villages, and this little district is situated at the inland extremity of the fine valley in which we now are, a few miles to the north or north-west of these remains. We see at Mílata how the name of an ancient city is sometimes transferred to a modern village, existing not on its site but in its neighbourhood[1]; and thus we may suppose that here the name of the ancient city has been transferred to the small district, and therefore consult the ancient authors with an expectation of finding something to prove that the city of Kantanos existed on this spot.

The Peutinger table gives us, as the distance from the western Kisamos to Kantanos, twenty-four miles: this agrees perfectly well with the distance between

[1] See Vol. I. p. 269. not. 48. and compare Vol. I. p. 63. and above, p. 78.

Kísamo-kastéli and these ruins near Khádhros. We know, on the same excellent authority, that from Kantanos to Lissos was sixteen miles: here too we see, that reckoning four miles from this place to Sélino-kastéli, and following our route thence to the ruins at Hághio Kýrko, we obtain a distance which again agrees perfectly with that of the table.

The city Kantanos is mentioned in Stephanus of Byzantium[2], and is also found in Hierocles's list of Cretan cities, where it is called Kantania, and is placed between Kisamos and Elyros. It was a bishop's see under the Byzantine Emperors, and when the Venetians obtained the island they established a Latin bishop here as in every other diocese. A letter from Gregory XI. written from Avignon in April 1375, to the bishop of Cantanus, is preserved[3]: as well as a diploma by which Clement VI. in 1346, appointed a new bishop to the see.

Spaniáko, where we arrive at three P.M. is about a mile to the south-west of the remains. We have now nearly returned to Sélino-kastéli, which is only about three miles further southward. Between the site of Kantanos and the village of Spaniáko, I fall in with a Mohammedan who has been at London, as one of the attendants of a Turkish ambassador, and who, on other occasions, has visited Leghorn and Marseilles. He is eloquent in his praise of Europe, and of the material comforts and luxuries which are there so common, while here they are entirely unknown. I learn that there are ancient remains, near a village called Vlithiás, a little to the north; and also, from a Sfakian, that at Anópolis in Sfakiá, there are very considerable remains of walls, the stones of which are of an enormous size.

[2] STEPHANUS BYZANTINUS says, in a passage the corruptness of which it is as easy to see, as it is difficult to find out the true reading: Κάντανος, πόλις Κρήτης, ὡς Ξενίων ἐν περιορίου Κρητικοῦ τόπῳ. τὸ ἐθνικὸν Καντάνιος.

[3] CORNELIUS, Creta Sacra, Vol. II. pp. 55 and 168.

I also learn, in conversing with the Mohammedans of this village, that, to the south-east of Haghía Iréne, is a thólos or vaulted apartment, six or seven feet high, near thirty feet long, and ten or twelve feet wide. There are remains of what would seem to have been two towers, half a mile to the south of Spaniáko: they were visited by my companion, while I staid conversing with the Mohammedans of the village.

WALL OF RUIN NEAR VLITHIAS.

CHAPTER XXX.

RUIN OF AN ANCIENT TOWER OR SEPULCHRE NEAR VLITHIAS. ARRIVAL AT THE HAMLET OF VLITHIAS. DEPOPULATION CAUSED BY THE LATE WAR. HEAR OF ANOTHER ANCIENT SITE.

April 30, *continued.*

A GUIDE accompanies us from the village of Spaniáko to the ruins found, about two miles off, near Vlithiás. A natural rocky elevation is here surmounted by very ancient walling, which affords a beautiful specimen[1] of what is commonly termed the second cyclopian style, and would seem to have been a sepulchre. The monument being round, and of a very solid and beautiful style of construction, reminds me of Caecilia Metella's tomb in the neighbourhood of Rome: but they differ greatly in their age; for this monument was undoubtedly in existence long before the earliest beginnings of

[1] The portion engraved is 16 feet 3 inches long, and about 6 feet high.

Rome and the Romans. The internal diameter of the tomb is about fourteen feet, and the thickness of the walls four feet. The sketch is taken from the east, on which side the rock is about twelve feet high: on the west one can just walk round it, at eighteen or twenty feet below its summit, (and above this path I observed steps cut in the solid rock,) but there is still a considerable precipitous descent even below this path.

We arrive soon after sunset in the little village of Vlithiás, the only male inhabitant of which is a young Mohammedan, in whose house we took up our abode. The rest are all widows. In many places in Crete the number of widows is large: and in one village of Lassíthi they actually form the entire population, as is so nearly the case here. These are striking instances of the depopulating and exterminating character of the late war, so different from those carried on between civilized nations, with whom the effects on population, as even under the most rigorous conscriptions of Napoleon, are almost imperceptible.

This almost total extermination of the male inhabitants in some parts of Crete, and the general aspect of its villages, all of which may be said to be in ruins, recall to mind the scene which was presented to the eyes of Demosthenes, as he travelled through Phocis[2].

Having obtained a young lamb from a Mohammedan at Spaniáko, for the very reasonable price of seven Turkish piastres, somewhere about eighteen pence in English money, we fared better this evening than was usual with us.

I had also the satisfaction of learning that the distance is not very great to the ancient site, near which I passed, without visiting it, between Pelekánas and Kástel-sélino; and, having reaped so rich a harvest

[2] DEMOSTHENES, p. 361. Ἐξ ἀνάγκης ἦν ὁρᾶν ἡμῖν πάντα ταῦτα, οἰκίας κατεσκαμμένας, τείχη περιῃρημένα, χώραν ἔρημον τῶν ἐν ἡλικίᾳ, γύναια δὲ καὶ παιδάρια ὀλίγα.

of discovery during the last few days, I can now afford to lose a day in an attempt to glean something, even if the field should prove barren: I therefore determine to start early to-morrow morning to examine the spot.

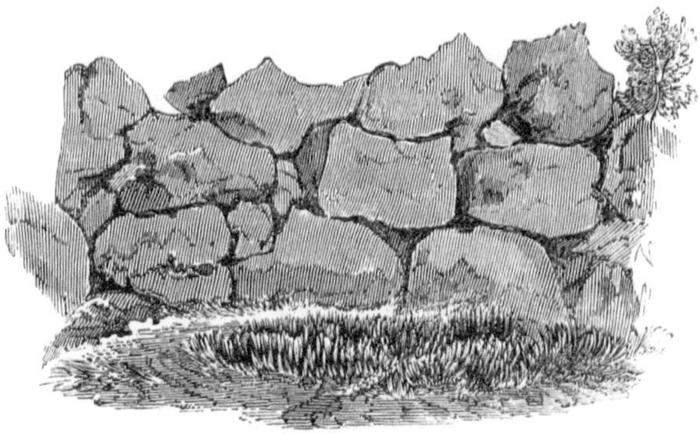

WALLS OF KALAMYDE.

CHAPTER XXXI.

VISIT THE SITE OF KALAMYDES. A GENERALLY PREVALENT SUPERSTITION. POLITENESS OF MANIAS IN SPEAKING TO A MOHAMMEDAN. THE LIGHT IN WHICH JEWS ARE HELD BY ORIENTAL CHRISTIANS.

May 1.

I LEAVE Vlithiás at half-past six, and I find the descent steep and the road extremely bad down to the river, which is even now a considerable stream: after passing it I begin to ascend, winding in a south-westerly direction along and up the side of the mountain, which I soon cross, and descend at a place called Kontokynéghi[1], into a valley parallel or nearly so to that of Kántanos. I reach Kontokynéghi at twenty minutes past seven: it is one of the most sequestered and quiet spots imaginable, entirely shut in by the lofty rocky hills of the valley, and half buried in the thick grove of tall olive-trees, carobs and almonds, in the midst of which it is situated. This is one of the many spots in Crete

[1] Κοντοκυνῆγι.

which, if one could but be surrounded by some of the ordinary comforts of European life, would be a delightful refuge from the tumult and anxieties of the world: a fitting spot wherein,

Ducere sollicitae jucunda oblivia vitae.

After passing the river which flows along this valley of Kontokynéghi, I ascend the ridge which separates it from that of Pelekánas, and soon cross the road by which we had traversed the ridge on our way from Pelekánas to Sélino-kastéli: a few minutes more and I arrive at a church of Hághios Antónios, after passing which I notice several tombs hewn out of the solid rock, a never-failing indication that an ancient site exists somewhere near. About a mile from the church of Hághios Antónios I arrive at another of Hághios Gheórghios and also at the ancient site of which I am in search. It is situated on the summit of the ridge between the two valleys: on the western and south-western sides of the city the walls may be traced for three or four hundred paces: on the east I could only observe them for about one hundred paces, while on the south the ridge narrows, and the wall, adapting itself to the natural features of the hill, has not a length of more than fifteen or twenty paces. The whole seeming circuit of these walls cannot much exceed half a mile: still, possibly, the city might extend further to the north. Foundations of the walls of buildings are seen to the south of the church of Hághios Gheórghios. Of the walls which remain the style is ancient though the construction is not very massive: the chisel has not been used for any of the stones: the sizes of most of them are pretty nearly the same, and they are all polygonal. The thickness of the wall is about four feet.

I suppose these to be remains of Kalamyde, of which the Coast-describer[2] says that it was to the west

[2] Anonymi Stadiasm. in Geogr. Graec. Min. Tom. II. p. 496. See also above, p. 86.

of Lissos, and thirty stades distant from Kriu-metopon. I know of no other city mentioned in any ancient writer, which we should be authorized to place here; and the site agrees perfectly with what we learn of Kalamyde from the author in question, who alone has recorded the name of this city.

I returned to Vlithiás by half-past ten.

Some days ago, while I was washing, I asked Captain Maniás to reach me a piece of soap, which was lying near him. He placed it at some distance from me, and told me that no motive could ever induce him to put it directly into my hands. The superstition, that when one person so gives soap to another, it will wash away their friendship, is generally diffused in Greece and Turkey[3].

I could not but notice Maniás's politeness, when, addressing our Mohammedan host at Vlithiás, he spoke of "those animals which have bristles on their backs," and carefully avoided even the name of the unclean pig. In a similar manner a Greek will apologize to any one before whom he may mention a *Jew*[4].

[3] Sir JOHN HOBHOUSE, Travels in Albania, Vol. I. p. 33. observes, "The captain would not give the soap into my hands, though I was sitting close to him, but put it on the ground within an inch of me.—I found that in Turkey there is a very prevalent superstition against giving soap into another's hands: they think it will wash away love."

[4] This observation is made by HARTLEY, Researches in Greece and the Levant, p. 206. "It is impossible to give an adequate idea of the contempt in which the Jews are held by the Greeks. The style in which they sometimes speak of them may, in part, illustrate this assertion. When the Greeks have to mention swine, and some other objects which they deem particularly offensive, they usually introduce the expression, *μὲ συμπάθειαν*, *begging your pardon*, as a duty of politeness to the persons present. A similar mode of speaking is often adopted, when there is occasion to introduce the mention of a Jew:—'I was walking along the street, and I met, *begging your pardon*, a Jew!'"

MONUMENT AT VLITHIAS.

CHAPTER XXXII.

DEPARTURE FROM VLITHIAS. SFAKIAN WOMEN. THE OLIVE-TREES OF SELINO. A BEAUTIFUL FEMALE PEASANT. VIEW OF THE NORTHERN AND SOUTHERN SEAS. ARRIVAL AT ERGASTERI. FISCAL INNOVATIONS OF MEHMET-ALI-PASHA. SONG ON THE DEATH OF THE CHRISTIAN CHIEFTAIN TZELEPES. A RUMELIOTE SONG ON THE DEFENCE OF MESOLONGHI. SONG ON THE DEATH OF KHADJI-MIKHALI.

May 1, *continued.*

We set out from Vlithiás at noon. Not a breath of air was stirring either yesterday or to-day, so that the heat is very considerable. About one o'clock, on entering the straggling village of Kakotíkhi, I was struck with such an apparition of female beauty, as, when once seen, can never be forgotten. This Cretan maiden's features were certainly more heavenly than I had ever seen in any "mortal mixture of earth's mould." She professed ignorance on the subject of my inquiries, and at the same time drew her veil half over her face. So strange an event as her being addressed by a European traveller at once called up

The embarrassed look of shy distress
And maidenly shamefacedness.

Very few moments elapsed before several other more matronly persons approached, and, in a few minutes, all the female population of the village was congregated around us. Their language at once told me that they were all from Sfakiá, and I learnt that they come down here during the winter to gather olives, and that they have been so employed for five or six months. They, however, return to their mountain-village Hághio Yánni to celebrate Easter Sunday, a day on which all Greeks think it almost a sin not to be at their homes. I observed to some of them, that the olive-trees, which vary greatly in size in different parts of Crete, are very fine in Sélino: the most talkative old dame of the group, whose tongue scarcely rests for an instant during my stay with them, and whose language, like that of all her companions, was uniformly characterized by the most strongly marked peculiarities of her native dialect, replied, "Yes, yes! and the oil too is excellent, and abounds in the whole district of Sélino, little as we have of it in Sfakiá[1]." It is natural that the women, who dwell so much at home, and associate so little with any one except the inhabitants of their respective villages, should preserve more purely than the men the idiomatic peculiarities of their native mountains. One of the women began by observing, "You are only two men, and therefore we dont fear you, for we are many." Had we, the travellers, been a more numerous party, or had

[1] The oil, λάδι, was called ῥάδι, and was said to be καρὸ (καλό).

they been fewer, it is plain that the beautiful young girl whom I had first addressed, a native of the Sfakián village of Hághios Ioánnes or Yánni, would have run away, and the others would not have ventured to come out of their houses. The men, who are seldom found at home except early in the morning and towards evening, were all busied in the fields or were gone to the city. On leaving this singular group, with whom I staid some time, I confess I "cast a longing lingering look behind" on her whose features, once seen, could hardly ever be forgotten:

Cose appariscon nello suo aspetto
Che mostran dei piacer di Paradiso—
Elle soverchian lo nostro intelletto:
Come raggio di sole un fragil viso:
E perch' io non la posso mirar fiso,
Mi convien contentar di dirne poco!

In three quarters of an hour, continuing to follow the course of this fine valley, we arrive at Plemmelianá, one of the small villages which are comprised under the common name of Kántanos. This place is on the western bank of the river, both sides of which are thickly covered with fine platanes: we see vines twining round them, as in other parts of the island, to a height of thirty or forty feet. I remain here to have all our steeds shod: I had despatched Maniás before us this morning with my own horse; and now the same operation was performed on the mules. There are but few villages in which one can find a blacksmith, and if a horse or mule happens to lose a shoe, the traveller in Crete will, in all probability, be obliged to continue his journey for two or three days, before he finds the means of replacing it.

We leave this village at twenty minutes past three, and follow the river, with its platanes, to which I see that ivy attaches itself still more closely than the vines, for about half a mile, when we pass, for a few minutes, through a double hedgerow of myrtles, which is succeeded by a plantation of olive-trees. At length we arrive at the village of Kufalatós, having on the way passed another of the hamlets of Kántanos, where an Albanian bulúkbashi was very anxious to see my teskeré from the Pashá, and where I was shewn what would seem to have been an ancient cistern. At Kufalatós we began to ascend, our course continuing, as it has been for some time, nearly due east. This ascent lasts for no less than three quarters of an hour. On reaching the summit of the mountain we have a wide view, extending northward to Cape Spádha and the whole bay of Kísamos, and southward to the African sea, including also the mountains of Sélino, and spreading right over to those of the Grabúsian promontory. On this summit was the church of the cross, a post where the Mohammedans established themselves in the early part of the war, and from which the Christians had great difficulty in driving them. We see, on looking back, the head of the valley, in which is situated the little district of Kántanos, nearly entirely surrounded by mountains: it was in this district that the Mohammedans, of nearly the whole eparkhía of Sélino, were attacked by the Christians in the second year of the war.

We descend till six, when we pass, to the left, the village of Lukianá, and commence a slight ascent, pass the village of Skáfi, and, at about twenty-five minutes past six, arrive at Ergastéri[2], with the grammatikós of which village we take up our quarters for the night.

This evening the novelty of the sight of European travellers draws together an assembly of ten or twelve peasants at our host's house, and some of them were from other villages. I succeeded, in a short time, in quieting the fears which seized on them, at first, when I inquired about the number of houses in this and the

[2] Ἐργαστήρι, more commonly called Ἀργαστήρι.

neighbouring villages. One of them, a Mohammedan, as I learnt afterwards, immediately whispered to my host, saying, "Dont tell him: he must be sent by the Pashá, and it can only be to put a tax on each house, that he wants to know how many there are." I mention this not as a peculiar incident, but, as to the distrust entertained of me at first when I arrive any where, a daily one: so great was it that I had to talk at least half an hour or an hour every day with the people, before I could commence asking questions on any subject on which I wished to obtain information. If I asked a question before gaining their confidence, I could never obtain an answer: nobody knew any thing.

I was able this evening, from the goodly assembly of villagers here met, to verify the account of the villages and population of Sélino, which I had made in the last few days; and found, as I had hoped, that in every instance its correctness was nearly perfect. I listen to several songs, in which the events of the war are celebrated: one on the death of Tzelepés, at the spot we passed this afternoon. At last the villagers begin to unburthen themselves of their grievances: to speak to me of the proceedings of the government: of the merciless changes in the system of taxation: of the duties on the leather for their boots, of the tithe of their almonds and silk, of the piastre for each místato of their wine, and of the various other extortions, which, they say, make them all, Christians and Mohammedans, long for the Frank, that is for some European power, instead of the Egyptians. One of them said, facetiously, "I ought to prefer the Turk, for I have got some fine old Frank olive-trees, which I suppose the Franks will make me pay for, if they come back again." It is probable enough that his olive-trees existed for centuries before the Venetians lost the island.

Several of the villagers said, that they heard yesterday of my being at Rhodhováni, but did not believe it. I asked why: they answered, "because it was said, that you wrote down a great deal about the Hellenic remains: which we did not understand; but, now that we see it with our own eyes, we believe it." Perhaps they may be pardoned for not having believed in an event, which is certainly contrary to the ordinary course of nature in Sélino, until they were convinced of its occurrence by the evidence of their own senses, when they saw, heard and talked with me.

The party did not break up till near midnight.

The following is part of the song on the death of Tzelepés, of which I have just spoken.

[3] Ill luck the Sunday e'er betide,
And perish the Tuesday,
When far-famed Tzelepés came here
To engage in hostile fray.

Gaúres 'twas he sought to find,
Up at the cross's height,
As with him he his cannon dragged,
With Moslem foes to fight.

* * *

[3] Νὰ ἔχῃ βούλισιν ἡ κυριακὴ
καὶ νὰ ῥαΐσῃ ἡ τρίτη,
ὥντεν ἐκεῖνος ὁ Τζελεπὴς
ἦλθε νὰ πολεμήσῃ.

Καὶ τὸν Γαοῦρι ἐγύρευγε
εἰς τοῦ Σταυροῦ τὸ σπῆθι
καὶ τὸ κανώνι τοῦ ἔσυρνε
νὰ τῶνι πολεμήσῃ.

* * *

The Tzelepés his sword unsheath'd,
The roof to perforate,
In order on them fire to cast,
That all might meet their fate.

Then Zunalákes at him aimed,
And did his weapon rest
So that its ball was buried deep
In Tzelepés's breast:

It entered deep beneath his breast,
And to his throat it ran;
Nor e'er till then the Tzelepés
Was heard to cry, Amán!

* * *

Τὸ μαχαιράκι τοῦ 'συρε
τὸ δῶμα νὰ τρυπήσῃ
νὰ τῶνι βάνῃ καὶ φωθιά
καὶ ὅλους νὰ τζὴ κεντήσῃ.

Δῶμα *tectum*. This sense of the word is unaccountably omitted by Passow, though it occurs, not only in the Greek Scriptures, Old and New, but also in Josephus, Herodian, and Aesop.

Τὸ Ζουναλάκι τοῦ 'παιξε,
καὶ τοῦ 'στεκε τζελέκι,
γιὰ τὶ τούτην ἐξάμμωσε
τοῦ Τζελεπὴ στὸ μπέτι.

Στὸ μπέτι τοῦ τὴν ἐξάμμωσε,
στὸ στόμα τοῦ τὴ βάνει,
καὶ τότε ἐφωνάξενι
ὁ Τζελεπὴς, ἀμάνι!

'Αμάνι, *Amán*, (ἔλεος or συγγνώμη in ancient Greek, in English *Mercy*,) like several other words of this song, is of Turkish origin.

"Stand on thy feet O Tzelepés,
That we the assault may push:
Perhaps the Moslems e'en may dare
From out the house to rush."

* * *

The Christians went and buried him
At Sfakiá's chief town,
Because e'en sweet as violet's flower
Was Tzelepés' renown.

At length the Moslems ventured forth
As evening's shades drew near,
Nor knew they that the Tzelepés
Was stretched upon his bier.

* * *

Σήκωσε ἀπάνω Τζελεπής
νὰ κάμωμεν γεροῦσι,
ἐμπορετὸ νὰ πορίσουσι
ἀπὸ τὸ σπῆθι οἱ Τοῦρκοι.

καὶ πάγουν νὰ τόνι θάψουνι
στῶν Σφακιανῶν τὴν χώρα
διατὶ τὸν ἔλεγαν Τζελεπή,
καὶ ἦτον καὶ ὡσὰν ἡ βιόλα.

καὶ ἀργὰ ὡσὰν ἐμούδισε
πορίζουνι καὶ οἱ Τοῦρκοι
δὲν 'ξεύρουνι τὸν Τζελεπὴ
ὅπως εἶναι στὸ τὸ ταμπούτι.

Πορίζω is of very common occurrence in the language of Crete, but is not used, as far as I am aware, in any other part of Greece or Turkey.

Some years later, during the protracted siege of Mesolónghi, Mustafá-pashá was encamped in the district of Mesághia[4]. In order to produce a belief in Crete, that Mesolónghi, which the Greeks so heroically defended, had fallen, the Pashá more than once gave out, that he had received positive intelligence to that effect, and also fired salutes in honour of the event[5]. Before I left the island, I heard, from a Rumelióte, a song grounded on similar rumours, which were spread, by the Turks and their friends, elsewhere as well as in Crete. It will be observed that there are no rhymes in the verses of the original: in this respect it resembles the other songs of continental Greece[6].

[7]Would that on high I could ascend,
And like a bird could fly,
To gaze on Mesolónghi's walls
In distant Rúmeli.

[4] Mentioned above, p. 60.

[5] CORRESPONDANCE DU CONSUL DE FRANCE: "*Le* 30 *Decembre*, 1825. Mustafa bey—qui se trouve en ce moment à Kissamos, avait reçu une lettre qui lui annonçait—la nouvelle positive de la prise de Missolonghi. Il avait desuite fait tirer le canon, et avait donné l'ordre au Mousselim de la Canée de faire aussi tirer le canon dans cette forteresse, et à la Sude." Again, "*Le* 17 *Fevrier*, 1826. Il est evident dès lors que Mustafa bey, en faisant tirer, dans les forteresses de l'île de Candie, le canon pour annoncer que Missolonghi était prise, avait pour but d'intimider les bandes de Grecs insurgés, qui parcourent les montagnes de cette isle, et espérait, par ce moyen, obtenir leur soumission." The fall of Mesolónghi was not known at Khaniá till the beginning of May.

[6] FAURIEL, Discours préliminaire, p. CXX.

[7]Νὰ 'μοῦν πουλὶ νὰ πέτανα
νὰ πήγανα στοῦ ὑψηλοῦ
νὰ θώρουνα, στὴν 'Ρούμελη,
τὸ δόλιον Μεσολόγγι·

This πουλὶ is of perpetual occurrence in the songs of Northern Greece.

To see with all the Moslem host
And four Pashás its fight,
While balls like drops of rain descend,
And bombs like hail alight.

To see there too the light tuféks
Like sands on the sea-shore;
They said that it surrendered had,
And would contend no more:

They said that Mesolónghi had
Submitted to the foe,
While still in war alone she seeks
Her bravery to shew.

The death of every chieftain who fell during the war is thus celebrated in popular songs[8]. I have heard several, in different parts of the island, in which Khadjí Mikháli's expedition to Crete, and his death near Fránko-Kástello in Sfakiá, were celebrated. The following are parts of one of the most interesting of them.

πῶς πολεμᾷ μὲ τὴν Τουρκιά,
μὲ τέσσαρους Πασάδαις,
πέφτουν αἱ μπάλλαις σὰ βροχὴ,
αἱ μπόμπαις σὰ χαλάζη·
κ' αὐτὰ τὰ λιανὰ τουφέκια
σὰν ἄμμο τῆς θαλάσσης·
τὸ λέγαν νὰ παράδοθη,
τὸ λέγαν νὰ προσκυνήσῃ·
κ' αὐτὸ γυρεύει πόλεμον
θέλει νὰ πολεμήσῃ.

[8] See those on the deaths of Búzo-Márko, Glemédh-Alí and Theódhoros, in Vol. I. pp. 78. 110. and 165.

[9] On every holy festival,
On Sunday, Easter-day,
Listen to what I now will tell,
Khadjí Mikháli's fray.

When the Grabúsians did write
Of their affairs the sum,
And it to the Moréa sent
That the Khadjí might come;

They also wrote to the Khadjí,
That ancient Moreíte,
That he should gather steeds of war,
And go down into Crete.

While at Anápli he collects
Horsemen full sixty-five,
That with them he to Crete may go
And with the Egyptians strive.

[9] Πᾶσα λαμπρὴ καὶ κυριακὴ
πᾶσ' ἑορτὴ μεγάλη
ἀφρουκαθῆτε νὰ σᾶς 'πῶ
διὰ τὸν Χατζὴ Μιχάλη.

'Εκάθσανι οἱ Γραμπουσιανοί
καὶ γράψαν ἀρζουχάλι·
στὸν Μορεὰν τὸ πέψανι
νὰ 'λθῇ ὁ Χατζὴ Μιχάλη.

Καὶ πέψαν γρᾶμμα τοῦ Χατζῆ
τοῦ παλαιοῦ Μορεΐτη,
νὰ περμαζώξῃ ἄλογα,
νὰ καταβῇ στὴν Κρήτη.

Στ' 'Ανάπλι μονομέρισε
ἐξῆντα πέντε ἀτλίδαις
στὴν Κρήτην διὰ νὰ καταβῇ
ἀποῦ 'ναι Μεσιρλίδαις.

His men, when they collected were,
  He embark'd in ships of war:
He had selected Rumeliótes
  Each a brave pallikár.

He went and landed all his men
  Upon the desert shore;
And asked the Grabúsians
  If they had powder in store.

"Powder have we, and balls in store
  Wherewith the war to wage;
We only long thy steeds to see
  On Creta's plains engage."

Yet, still, he ventured not to trust,
  But to his boats did turn,
And landed at the port Lutró,
  The truth that he might learn.

'Απεῖ τζὴ μονομέρισε
  τζὴ ἔβαλε στὰ καράβια
ἐδιάλεξεν τζὴ 'Ρούμελεις
  τὰ ὤμορφα παλλικάρια.

Πάγει καὶ 'ξεβαρκάρε τζὴ
  στὴν ἐρήμην Γραμποῦσα,
κ' ἠρώτα τζὴ Γραμπουσιανούς,
  ἂν ἔχουνι μπαρούθια.

"'Ημεῖς μπαρούθια ἔχομεν
  βόλια νὰ πολεμοῦμεν,
ἄλογα μόνον θέλομεν
  στὴν στερεὰν νὰ βγοῦνι."

Μὰ πάλιν δὲν ἐπίστεψε
  μὰ 'μπῆκε στὰ καΐκια,
νὰ 'ξεβαρκάρῃ στὸ Λουτρό
  νὰ μάθῃ τὴν ἀλήθεια.

There, on the shore, he Sfakians found,
Men all of tried valoúr,
And who, for every feat of war,
Were highly in honoúr.

"Come hither all ye Sfakians,
Come also ye Rhizítes,
That we may go and make revolt
E'en the Katomerítes.

Come hither all ye Sfakians,
Ye who are warriors hight,
Leave all your flocks, that we may rush
On the Moslem host in fight."

And the Pashá this news did find
Most grievous for to hear,
At Kástro, and at Rhíthymnos,
Where now it met each ear.

Ευρίσκει έκεῖ τζὴ Σφακιανούς
ἀποῦ ἦτον ἀνδρειομένοι,
κ' ἦτονι, κ' εἰς τὸν πόλεμον,
περίσσια τιμημένοι.

"'Ελᾶσθε ὑσεῖς οἱ Σφακιανοί
καὶ νὰ 'λθουν κ' οἱ 'Ριζῖταις,
νὰ πάμε νὰ σηκώσωμεν
καὶ τζὴ Κατωμερῖταις.

'Ελᾶσθε ὑσεῖς οἱ Σφακιανοί
ἀποῦ εἶσθε παλλικάρια,
νὰ πολεμοῦμεν τὴν Τουρκιά,
κ' ἀφεῖσθε τὰ κοπάδια."

Καὶ ὁ Πασιάς, ὡς τ' ἄκουσε
πολλὰ τοῦ βαρυφάνει—
στὸ Κάστρο, καὶ στὸ 'Ρίθυμνος
τὸ μουκιαρέμι φθάνει.

"Assemble all ye bravest men
  Of Islam's faith, to-day:
That we may make this stranger flee
  To the hills from out the fray.

For sure he'll perish midst the hills,
  Or in the sea will fall;
Or we shall kill the man who comes
  As if to meet my ball."

This speech when his Kiaias heard,
  He turned, and said, "From war
This man will never flee, Pashá,
  For he's a pallikár.

This is not a Lazópulo,
  To the mountain-heights who springs,
But comes from out of Rúmeli,
  And with him warriors brings.

"Περμαζωκτεῖτε τῆς Τουρκιάς
  πρόβολα παλλικάρια,
νὰ πάμε νὰ τόνι σφίξωμεν
  νὰ πιάσῃ τὴν μαδάρα.

"Η στὴν μαδάρα νὰ χαθῇ,
  ἤ στὸ γιαλὸ νὰ πέσῃ,
ἤ νὰ τόνι σκοτώσωμεν,
  γιατὶ ἦλθε σὰν πεσκέσι."

Μὰ ὁ Κιαιάς του τὸ δρικᾷ,
  γυρίζει, καὶ τὸ *κάνει,
"Δὲν φεύγει, Μουσταφά-πασιά,
  γιατὶ 'ναι παλλικάρι.

Δὲν εἶναι αὐτὸ Λαζόπουλος,
  νὰ πιάσῃ τὴν μαδάρα,
μόνον εἶναι 'πὸ τὴν 'Ρούμελη,
  καὶ σύρνει παλλικάρια.

With him Bulgarian youths he counts,
  Horsemen of worth approved:
And us full surely they'll destroy,
  Howe'er we may be moved."

"Such fears, Kiaia, know not I!
  Him, with his horsemen good,
Like herbs in salad I'll devour,
  Or anchovies, for food."

*   *   *

The self-same hour that the Pashá
  The unwelcome truth did ken,
Forth from the walls of Khaniá,
  He marched with all his men.

*   *   *

Καὶ σύρνει Βουλγαρόπουλα,
  ἀτλίδαις τιμημένοι,
καὶ θὰ μᾶσι, σκοτώσουσι,
  καὶ ἂς εἴμεσθα γνοιασμένοι."

"Ἔτσι φοβοῦμαι 'γώ, Κιαιά;
  μὲ τόσους καβαλλάρους
σὰν τὴν σαλάτα τρώγω τζή
  ὡς τρώνι τζὴ ψαρογάρους."

*   *   *

Καὶ ὁ Πασιὰς ὡς τ' ἤκουσε
  ἐκείνηνα τὴν ὥρα
μονομέρισε τζὴ ἄλλους του
  καὶ βιαίν' ἀπὸ τὴν χώρα.

*   *   *

The assemblage of all the Pashá's forces in the plain of Apokórona, is next described.

He bade the Kastrians too to advance,
  With the Rhithymniótes,
That they might all their forces join
  With those of the Khaniótes:

That when they came, and were conjoin'd,
  In single column good,
They might proceed, to take and sack
  Sfakiá, nor be withstood.

* * *

They march, and all their forces join
  At the Hellenic bridge,
And this the Christians also hear,
  And flee to the mountain's ridge.

The passage of the Pashá, through the gorge of Askýfo, into Sfakiá, is followed by a description of the consultation (*κοσούλτο*) held by the Rumeliótes, near the southern shore. The advice of Kyriakúles is men-

Μηνύουνι καὶ τῶν Καστρινῶν,
  νὰ 'λθουν, κ'οἱ 'Ριθυμνιώταις,
νὰ 'λθοῦνι νὰ 'νταμώσουνι
  μαζὴ μὲ τζὴ Χανιώταις·

Νὰ 'λθοῦνι νὰ 'νταμώσουνι,
  νὰ κάμουνι κολώνα,
νὰ πάγουνι νὰ πατήσουνι
  τῶν Σφακιανῶν τὴν χώρα.

Πάγουν καὶ μονομερίζουνι
  τζὴ 'Ελληνικαὶς καμάραις,
καὶ τὸ γροικοῦνι κ' οἱ 'Ρωμαιοί,
  καὶ πιάνουνι τζὴ μαδάραις.

tioned: and also the directions given by Khadjí Mikháli himself. At the end of the latter's speech his troops, with one voice, exhort him not to expose his person in the approaching contest.

His warriors then, both great and small,
  With voice united said,
"Share not, Khadjí, in this day's fight,
  Thou only art our head.

Thou only art the head, Khadjí,
  And refuge of thy host;
And, if the foe thy death should cause,
  Then all our lives are lost."

It is in vain that they thus try to dissuade him from taking an active part in the combat: he answers them as follows:

"As once into the world I came,
  So, any one can tell,
That I, like every other man,
  Must bid it, once, farewell.

Τὸ ἀσχέρι τὸν παρακαλεῖ,
  ὅλοι μικροὶ μεγάλοι,
"Χατζῆ μὴν μπῆς στὸν πόλεμον,
  διατὶ εἶσαι τὸ κεφάλι.

Διατὶ εἶσαι τὸ κεφάλι μὰς,
  καὶ διὰ *καταφυγήν μας,
κ' ἂν εἶναι νὰ σᾶς σκοτωσούν,
  χάνωμεν τὴν ζωήν μας."

"Μία φόλα ἐγεννήθηκα,
  μία φόλα θὰ 'ποθάνω,
καμμία νὰ τὸν ἀπαρνηθῶ
  τὸν κόσμον τὸν ἀπάνω.

And nobler 'tis that I should fall,
  Far nobler 'tis to die,
Than that my honour should sustain
  Disgrace in every eye.

Do ye then saddle quick my horse,
  Into the fight I'll speed:
I hear that the Pashá doth come,—
  To take him is my meed."

The Khadjí next describes the probable events of the conflict, and gives fitting directions to his followers:

And then he offered up a prayer,
  And aid from God did seek;
And next he took his own light sword
  And hung it round his neck.

Καλῄτερα νὰ σκοτωθῶ,
  καλλιά 'χω νὰ 'ποθάνω,
παρὰ νὰ 'χω τὴν ἐντροπή
  στὸν κόσμον τὸν ἀπάνω.

Σελλώσατέ μου τ' ἄλογο
  στὸν πόλεμον νὰ κάψω,
καὶ ἀκούω ὁ Πασιὰς ἀποῦ ἔρχεται
  νὰ πάγω νὰ τόνι πιάσω."

Νὰ κάψω: this verb is of constant use in Crete, and is, I think, peculiar to the Cretans.

Καὶ κάνει παρακάλεσιν
  δέησιν στὸ θεόν του,
καὶ πιάνει τ' ἀλαφρὸ σπαθί,
  κρεμνᾷ τὸ στὸ λαιμόν του.

Then, offering up another prayer,
  The cross's sign he made:
His pistols next he duly took
  And in his girdle laid.

Scarce had he vaulted in his seat,
  When wept his charger good;
And then he understood full well
  His death it did forbode.

He offered up another prayer,
  And in his saddle sate;
Then onward spurred his trusty steed
  From out the castle gate.

Καὶ κάνει παρακάλεσιν
  καὶ τὸ σταυρόν του κάνει,
καὶ πιάνει τὰ πιστόλια του,
  εἰς τὴν μέσην του τὰ βάνει.

Ὥντεν ἐκαβαλλίκευε
  ἔκλαιε τ' ἄλογό του·
καὶ τότεσα τὸ ἐγνώριστε
  πῶς εἶναι ὁ θάνατός του.

Of the truth of this incident I was assured by many other Cretan peasants, besides those from whom I learnt the song. The notion is full three thousand years old, and is beautifully expressed by HOMER, IL. XVII. 436.

Ὣς μένον ἀσφαλέως περικαλλέα δίφρον ἔχοντε,
οὔδει ἐνισκίμψαντε καρήατα· δάκρυα δέ σφι
θερμὰ κατὰ βλεφάρων χαμάδις ῥέε μυρομένοιϊν,
ἡνιόχοιο ποθῇ· θαλερὴ δὲ μιαίνετο χαίτη.

The Cretan poet, however, endues his steed not with mere human affections, but with foreknowledge of the fate which was about to befal his rider.

Ἔκανε παρακάλεσιν,
  στὴν σέλλαν τοῦ καθίζει·
δούδει βιτζιὰ τ' ἀλόγου του,
  εἰς τὴν πόρτα 'ξωπορίζει.

The song ends with an account of the affray, and of the vain prowess of the Khadjí; who soon paid the price of his temerity, although, for awhile, "the Mohammedan host trembled as it saw the war which he waged."

LAKI.

CHAPTER XXXIII.

MOSQUES IN THE RURAL DISTRICTS OF CRETE. A CATERPILLAR INJURES THE VINES IN SELINO. ANCIENT AND MODERN SUPERSTITIONS. PRODUCE OF AN OLIVE-TREE IN SELINO. SNOW NOT COMMON AT THE SEA LEVEL IN CRETE. RIDE TO LAKUS. AN ACCOUNT OF THE LAKIOTES. A PASSAGE IN THE HISTORY OF THE VENETIAN RULE IN CRETE. THE ANCIENT CTTY RHIZENIA. PREPARATIONS FOR EASTER SUNDAY.

May 2.

THE Cretan converts to Islamism have erected but few mosques out of the principal towns of the island. There are, however, three in the eparkhía of Sélino, which does not possess a single Greek monastery.

In this neighbourhood a small insect called Kampia[1] eats the young sprouts[2] of the vine about this time of

[1] The word used was κάμπια: the ancient name is κάμπη.
[2] Τοὺς βλαστούς.

the year. On account of its ravages last year, my host obtained only six místata of wine from his vineyard, which ought to have produced fifty místata; in Kísamos, the same quantity of ground, ten days' work[3] of a man, would produce nearly two hundred místata.

This caterpillar sometimes infested the vineyards of the ancients, as it now does those of the moderns in the district of Sélino. In order to free the vine from its ravages, various superstitious charms used to be practised, some of which were of a very extraordinary kind, and must remain veiled in the decent obscurity of a learned language[4]. In more recent times a supposed remedy consisted in summoning the offenders into a court of law. As soon as they were thus publicly called on, to answer for their depredations, it is gravely said that they at once abandoned the vineyard[5].

In the district of Sélino, where the olive-trees are very fine, nearly two trees are ordinarily required to produce a místaton of oil: sometimes three roots would be requisite. A village here producing one hundred

[3] Μεροκάματα, from ἡμεροκάματον.

[4] GEOPONICA, XII. 8. p. 858. AELIAN, N. A. VI. 36. PLINY, N. H. XVII. 47. COLUMELLA, Hort. 357.

At si nulla valet medicina repellere pestem,
Dardaniae veniant artes, nudataque plantas
Foemina, quae, justis tum demum operata juventae
Legibus, obscoeno manat pudibunda cruore;
Et resoluta sinus, resoluto moesta capillo,
Ter circum areolas et sepem ducitur horti.
Quae cum lustravit gradiens, mirabile visu,
Volvitur ad terram distorto corpore campe.

The κάμπη does not confine its attacks to the vine: see THEOPHRASTUS, H. P. III. 22. 3. p. 486. and ATHENAEUS, II. p. 55. e.

[5] LEO ALLATIUS, de quorund. Graec. Opinat. p. 177. Relatum mihi est, opportunissimum remedium esse, si illi, tanquam rei, per praeconem, aut libello scripto, per nomina in curiam ad judices citentur. *Sic enim vel illico evanescunt.* Si pergunt adferre noxam, repetita citatione proclamantur. Ubi pertinacius persistunt, in contumaces sententia pronunciatur. Sunt etiam, qui pro damno, jam accepto, et detrimento, testes examinant. Mirum esse dicunt, vermiculos, qui omnia tolerant, judicium non sustinere; sed repente, ne deprehensi poenas luant, eodem tempore diffugere, atque evanescere. See another extraordinary remedy against the poor kampe prescribed by Michael Psellus, and another practised by Tuscan monks, at pp. 178–179. of the same work.

místata would probably possess nearly two hundred roots of olives. I am told here, as in every other place where I have made enquiries, respecting the consumption of oil by each Cretan family, that it may be estimated at four okes a week, at the least. A mother will hardly give bread to her children without pouring them out some oil into a dish, that they may moisten the staff of life, and render it more savoury, before eating it. Oil is used with all kinds of vegetables, as well as in preparing every sort of meat and fish: in short it enters into every dish in Crete, and though all Greeks use a good deal of it, there is a much greater general consumption of it in this island than elsewhere. "The Cretans are used to it," say they, as I have already mentioned[6], and it is perhaps the only article in which no one ever thinks of stinting himself.

Even at this elevation, which, with mountains so much loftier in sight, does not seem very great, the snow usually lies several feet deep for some days in December or January. Last winter was unusually severe here: snow fell even on the island of Gávdha. The severity of the last winter was however general throughout the east.

After breakfasting on eggs, milk, bread and cheese, we start and reach Epáno-khorió in fifty minutes. Haghía Iréne is a mile to the north-east of Epáno-khorió. I hear at both these villages most unfavourable accounts of the *descent* by which, after reaching the elevated plain called 'Omalos, it is usual to enter Sfakiá, the people all doubting whether a horse like mine would be able to pass one part of the road called Xylóskalo[7] with safety; and agreeing that the *probabilities* were that he would miss his footing, and be dashed to pieces. About the mules no fears are entertained: "they are accustomed to the bad road[8]." Not wishing my horse

[6] Οἱ Κρητικοὶ εἶναι *μαθημένοι εἰς τὸ λάδι·* see Vol. I. p. 121.
[7] Τὸ *Ξυλόσκαλο*, a term which sufficiently explains its own meaning.
[8] Εἶναι *μαθημένα εἰς τὸ κακόβολον.*

to meet with any accident, I determined to go to Láki, and thence, if I shall hear no better accounts of the Xylóskalo, to traverse the Rhízoma, and part of the plain of Apokórona, and thus to enter Sfakiá by the ordinary pass of Askýfo.

Having come to this decision, I leave Haghía Iréne at a quarter past ten, and soon after meet three Greeks who are driving two mules and an ass all laden with manufactures of various kinds, brought from Sýra, and which they are now going to dispose of in the different villages hereabouts. At a quarter past eleven, we reach the summit of this long ascent, and again have a view of both the Cretan and African seas. The plain of 'Omalos is about five miles to the east: it is three or four miles wide and five or six long. As we descend on the north side of the mountain-chain, which we have just crossed, the Akrotéri, Cape Spádha and the whole gulf of Khaniá come into view. We halt from soon after twelve till near one, and my companion makes a sketch of the wide view which we have before us. At twenty minutes before three we reach Orthúni, and, in about an hour more, arrive at the considerable village of Láki.

The snow-clad mountains, visible from Láki, have each of them a particular name. To the south-east is Aliákes, to the south-west, or nearly so, is Agúzi, and between these two is Papalákos.

The Lakiótes acted a most conspicuous and honourable part in the war. This was partly owing to their active habits, caused by the mountainous nature of their country; but still more to their having been accustomed to the use of fire-arms, both for the chase on their neighbouring mountains, and for less innocent objects, which, in so wild and savage a state of society as that which existed in Crete, before the outbreaking of the Greek revolution, used constantly to offer themselves. In those times they used often to have petty wars with the Turks: that is, a Lakióte would cut off a Turk who had given him what he thought a just cause of offence, and, if the perpetrator of the crime was found out, the Pashá would come to the village with a sufficient force, and burn his house, &c. If he was not found out, a heavy fine was imposed on the village. Old Nikolúdhes, my host, the proestós or head-man of Láki, had his house burnt in this way no less than three times: it was also twice confiscated by the Pashá, along with all the property of his uncle and relations. In these petty wars Nikolúdhes received nine gun-shot wounds, three of which, since his vest and shirt were open, the day being very warm, I saw on his breast the moment I first spoke to him.

In the same good old times[9], if any one had an enemy and had also a Sfakian friend, he used to avail himself of the latter, in order to get the other put out of the way. It was simply necessary to intimate his wish, and to make at the same time a suitable present, and the Sfakian would at once watch for his opportunity, and soon give the desired proof of his friendship[10].

Krustoghérako in Sélino, and some villages in the immediate neighbourhood of Láki, were the scenes of atrocities, while Crete belonged to the Most Serene Republic, such as in all probability have never been perpetrated by any Pashá in any part of the Turkish empire. The following account of the proceedings of Venetian nobles, and of the public representatives of Venice in the island, is drawn from the manuscript account of a Venetian writer, who plainly approves of the atrocities which he describes[11].

[9] Singularly enough the Greeks almost always say, in speaking of the period before the insurrection, *εἰς τὸν καλὸν καιρόν.*

[10] Maniás admitted the truth of the Lakióte's relation, and defended the conduct of his countrymen: "for after all it is nothing when is is done for the sake of a friend"—*και πάντα, πάντα, διὰ ἕνα φίρον δὲν εἶναι τίποτες.*

[11] This manuscript is in the Raccolta Correr, and is numbered 766. Its title is "Racconto di varie cose, successe nel regno di Candia, dall' anno 1182 che si sono rubellati dalla devotione all' Imperio Greco, sino l'anno 1669 che restò al poter dell' Imperio Ottomano. Formato dal Sign[r.] Trivan Pub[co.] nodaro Ducale."

From the time of the great revolt in the year 1363[12], the island remained tranquil, under the Venetian government, till the beginning of the sixteenth century. At that time, however, the Greeks of Sélino, Sfakiá, and the Rhíza, including some villages situated almost on the very plain of Khaniá, united together and refused to obey the representatives of Venice. Their leaders were George Gadhanole, a native of Krustoghérako, the Pateröpuli of Sfakiá, and some other families of the Archontópuli, as they were called. Gadhanole was elected Rector (Rettore, Ῥετούρης) of these provinces. Each department of the new administration was filled by its proper officers, all appointed by the Rector. Thus duties and taxes were now paid, not to the Venetian, but to the Greek authorities; and two independent powers coexisted for some time in the island. At length the Greek Rector suddenly presented himself at the country house of Francesco Molini, a Venetian noble, in the neighbourhood of Khaniá, and asked his daughter in marriage for Petro, "the most beautiful and the bravest of all his sons[13];" and in whose favour the Rector declared his intention of resigning his office, on the celebration of the marriage. The alliance was agreed on; the "Rettore" gave his son a massive gold ring, and the betrothal at once took place. The youth kissed his future bride, and placed the ring on her finger[14]: the wedding was to be solemnized the next Sunday week, at the Venetian's country house, a few miles out of Khaniá. Molini was merely to send for a notary and a few friends[15], and Gadhanole, with his son, was to be accompanied by a train not exceeding five hundred men. The Greeks left the country-house

[12] See above, p. 53.

[13] TRIVAN, l. c. fol. 29. "E il più bello e bravo di tutti li altri miei figliuoli."

[14] "La baciò, e gli mise l' anelo in dito, baciandola pure anco il vecchio suo padre."

[15] "Tre o quattro gentilhuomini."

of the Venetian gentleman without for an instant suspecting that he meditated any foul play[16].

The following morning Molini visited the Governor of Khaniá, and obtained his promise of co-operation in an attempt to obtain, for the supposed indignity, a satisfaction "which might serve as an example to posterity[17]." In order, however, to prevent any suspicion of his good faith, Molini despatched tailors to his country-house, to prepare new dresses for the wedding, and also sent presents of fine cloth to his future son-in-law. During the next few days, the Governor of Khaniá assembled about a hundred and fifty horsemen, and seventeen hundred foot-soldiers, within the city.

On the day before the wedding, Molini went from Khaniá to his house at Alikianó, accompanied by about fifty of his friends, to be present at the marriage on the morrow. He gave orders for roasting a hundred sheep and oxen, and for making all due preparations to celebrate the nuptials with becoming splendour. The Greek Rector arrived, accompanied by about three hundred and fifty men and one hundred women, on the Sunday morning, and was delighted at all that he witnessed. He was received by Molini with every demonstration of kindness and affection. After the marriage ceremony, the day was naturally spent in festivity and rejoicing. The Greeks eat and drank, and danced and sang. The Venetian sedulously plied his guests with wine, and the intoxication which was pretended by the Venetian nobles present, all of whom were parties to the plot, really overcame the whole host of the unfortunate and too confiding Greeks. Some time after sun-set, a rocket thrown up at Khaniá gave notice, to the Venetians, of the approach of the troops to consummate their design. The Greeks, overpowered by wine and sleep, were all dispersed about the palace.

16 "Spensierati d' ogni altro sospetto."

17 "Che servirebbe d' esempio a' posteri."

As soon as the military force arrived, most of the destined victims were at once bound hand and foot, but were suffered to sleep on till near sun-rise. At day-break Molini and the Public Representative of the Most Serene Republic hung the Greek Rector, the unfortunate bridegroom, and one of his younger brothers. Of the family of the Musúri three were shot, and the rest hanged on trees. Of the Kondí, sixteen were present, eight of whom the Venetians hung, and sent the others to the galleys in chains. All the rest of the ill-fated prisoners were divided into four parties: not, however, with any intention, on the part of their ferocious and treacherous foes, of mitigating the penalty to be exacted from them, for an equally merciless fate awaited them all. The Venetians hung the first division of them at the gate of Khaniá, the second at Krustoghérako, which village, the birth-place of Gadhanole, was also razed to the ground; the third at the castle of Apokórona, and the fourth on the mountains between Láki and Thériso, above Mesklá, to which village Gadhanole had removed from Krustoghérako.

"Thus," says the Venetian Chronicler, who proves himself a worthy reciter of such barbarities of his countrymen, "they were annihilated, and all men who were faithful and devoted to God and their Prince were solaced and consoled[18]."

The Senate at Venice, however, did not think the consolation it thus received sufficient; but elected one Cavalli as Proveditor, giving him the fullest authorities for extirpating the seditious Greeks[19]. In executing these instructions he marched out of Khaniá a little before midnight, and surprised the village of Fotigniacó, near Murniés, and about four miles from Khaniá, accompanied by all his troops. They surrounded the place, and dragged all its inhabitants, men, women and children,

18 "Et li huomini fideli et divoti di Dio e del loro Prencipe solevati et consolati."

19 "Per l' estirpazione degli huomini seditiosi."

out of their houses. They then set on fire every dwelling in the village[20], and, at day-break, the orders of the Venetian Senate began to be fully accomplished. Twelve of the Greek Primates[21] were hanged, and, "to cause still greater fear," Capelli sought out the pregnant wives of four of the principal persons of the place, and "cutting open their bodies with large knives," tore forth their unborn children; "an act which truly inspired very great terror throughout the whole district[22]." This was the deliberate proceeding of a high public functionary of the Venetian Senate! "It is true," adds the Chronicler, "their crimes merited a severer punishment, and such also followed[23]." For all the captives were conducted into the city, with their families, and great numbers of them were there put to death[24]: the rest "were transported from Crete to 'the three islands[25],' so that the wicked race was thus eradicated." Five or six individuals alone escaped, at the moment when Cavalli arrived, like a midnight murderer, and found refuge in the villages of Murniés and Kertomádhes[26].

Cavalli, after so satisfactory a commencement, pressed onward in his course, and next required all the Greeks of Kástel-Fránko, Apokórona, Sfakiá, Sélino, and Kísamos, "to appear in the city, and make their sub-

[20] It contained four hundred inhabitants. During the following days the Venetians compelled the inhabitants of the neighbouring villages to remove every stone from the place, so that not a vestige of it was left.

[21] "Dodici dei più primati."

[22] "Per maggior spavento fece ritrovar quatro donne gravide, moglie de' Capurioni, le quali *con cortellazzi fece aprire il loro ventre, et cavatoli fuori i bambini—questo atto veramente fece grandissimo terore a tutto il contado.*"

[23] "Meritavano peggiò castigo, et tanto anco seguì."

[24] "Fece prima decapitar et appiccar una quantità di loro."

[25] "Nelle tre isole."

[26] I suppose these proceedings of Cavalli are alluded to by GIULIO DE' GARZONI, in his Report to the Senate, (see above, Vol. I. p. 31.) when he says, speaking of the Sfakians, "Si sono quelle genti alle volte mostrate disobedienti alla Serenità V., et hanno convenuto li magistrati *darli sacco et tagliarne a pezzi molti, et altri levare dalle proprie habitazioni,* come fece l'Eccellentissimo Cavalli, con giusta causa nota alla Serenità V."

mission." Some of them ventured to do so, but the leading families, and the mountaineers, could not trust themselves to the word or mercy of persons, by whom all ordinary principles of justice and humanity were so wholly disregarded. Their property was therefore declared confiscated, their lives forfeited, and a price was set on their heads. The life of any inhabitant of the several proscribed districts could only be redeemed on a condition, which is perhaps more revolting to the common feelings of mankind, than any other savage act of these unchristian rulers. Pardon was promised to any one of the proscribed, on condition that he produced, in Khaniá, "*the head of his father, or brother, or cousin, or nephew.*" This generous humanity was extended even further: each additional head which one man might bring into the city, would entitle him to demand from the governor the forfeited life of another relation. It is painful to learn, that the lives of many of the unfortunate objects of this barbarous and inhuman persecution, were thus redeemed by the deaths of their dearest relatives[27].

At length a priest, of the family of the Patéri-Zapa, entered the city, accompanied by his two sons, and by two of his brothers, each individual of the mournful party carrying in his hand a human head. We know, from the terms of the Venetian law, how near and dear those who had thus heroically died, must have been to the five survivors, whose lives could only be rendered secure by this dreadful sacrifice[28]. The wretched men placed their bleeding offerings before the Signor Cavalli, and the other representatives of Venice, and with the bitterest tears stated whose heads they were[29]. The facts

27 "In tal modo furono seguiti diversi accommodamenti, et non in altro modo."

28 Of the five heads, the first belonged to a son of the priest, the second to one of his brothers, the third to his son-in-law, and the fourth and fifth to sons of one of his brothers.

29 "Le gittò alla presenza del Sr. Cavalli, et delli altri Rappresentanti, et con amarissime lacrime rappresentava a quali fussero esse teste."

were duly established by witnesses: even the governor, who had been sent to Crete, in order "to extirpate" the seditious Greeks, was moved: and the law was at length abolished.

The sketch engraved at the head of this chapter suggests a good idea of Láki; the house of my host is near the little church, on the ridge on which part of the village stands.

Close as we are to the Sfakian frontier, which is only six or seven miles off, above Omalos, I do not find in the conversation of these Lakiótes the slightest trace of those peculiarities of dialect which characterize the Sfakians. The Lakiótes also differ from the inhabitants of the loftier mountains in their bodily appearance, for though perhaps as active looking as any men in the world, yet they approach not in the least the lofty stature and commanding aspect of the Sfakians.

Thériso, which is about as celebrated as Láki in the history of the early events of the insurrection, is four miles to the east. Láki and Thériso are the principal places of what is called the Rhíza[30]. The whole of the district containing the lower slopes of the White Mountains, and lying to the south of the more level parts of Apokórona and the district of Khaniá, is called by this name. Now there was an ancient Cretan city, Rhizenia, which is mentioned by Stephanus of Byzantium[31]. In a mountainous country like Crete, there are many places for which Rhiza or Rhizenia would be an appropriate name, and it is impossible to assign, with any confidence, the precise situation of this ancient city. At all events it may have been somewhere on this Rhíza, along the whole of which we find indications of the existence of only two cities, Lappa and Polichna.

[30] See above, Vol. I. p. 18.

[31] Stephanus Byzant. Ῥιζηνία, πόλις Κρήτης. ὁ πολίτης Ῥιζηνιάτης.

The eparkhía of Rhizó-kastron probably derived its name from the fact of its "Castello" being near the western foot of the mountains, which spread over a great part of its extent. The Rhíza of these mountains and the Kástron of the district were in the same neighbourhood, and Rhizó-kastron thus became the name, first of the fortress, and next of the eparkhía. I do not, however, feel disposed to place Rhizenia in the district of Rhizó-kastron, for the sites of Pyranthos and Asterusia, of Priansos and Inatos, sufficiently occupy the ground in that part of the island.

We find the Greeks of Láki already beginning to prepare for the festivities of Easter Sunday, which, with them, after so long and so severe a fast, is welcomed with feelings to which Protestant and even Catholic Christians are strangers. The severities of the endless Sarakosté or Lent have been very frequently endured by us, especially when it was rather late before we arrived at our resting-place for the night; and these privations, added to the fatigue of moving early every morning, and of being actively employed ten, twelve and sometimes fourteen hours daily, have at length made my companion so weak and unwell, that it seems prudent for him not to venture into the Sfakian mountains, where illness would be a very serious matter, since medical aid could be obtained only from Khaniá. It is therefore decided that he shall return to the city, leaving me to prosecute my journey, and to explore the unknown regions of the Sfakian mountains without him, but still accompanied by the faithful and talkative Maniás.

Wishing much to enter Sfakiá by Xylóskalo, I again make inquiries as to the practicability of the descent; and am assured by the villagers, that even mules ought to be habituated to such a road in order to be depended on. Not long since a Lakióte went with his mule for the first time: the poor beast started back on seeing the precipice, and, losing its footing, was precipitated to the bottom. Now, anxious as I was to see the wild scenery of this pass, I had no wish to meet with any such accident, and therefore, at last, determined to go round by the plain of Apokórona, and to enter the White Mountains by the common defile of Askýfo.

ALIDHAKI'S TOWER AT PROSNERO.

CHAPTER XXXIV.

DEPARTURE FROM LAKI. MESKLA, THERISO, DHRAKONA. EVENTS AT KERAMIA IN JUNE 1821. KAMPI. THE PASCHAL LAMB. RHAMNE, FRE, AND IPO. OLIVE-TREES OF APOKORONA. THE COMMON RETURN FOR THE INVESTMENT OF CAPITAL IN CRETE. EASTER-SUNDAY, WITH ITS FESTIVITIES, SALUTATIONS AND SUPERSTITIONS. SONG ON THE CAPTURE OF ADRIANOPLE BY THE MOHAMMEDANS. PYRGO OF ALIDHAKI AT PROSNERO. DEFILE OF ASKYFO. GREAT ROUT AND SLAUGHTER OF MOHAMMEDANS IN AUGUST, 1821. PLAIN OF ASKYFO. ARRIVAL AT ONE OF THE HAMLETS OF ASKYFO. PROCEEDINGS OF THE EGYPTIANS IN 1833.

May 3.

I LEAVE Láki at half past six, and, after a rugged descent, performed chiefly on foot, to Mesklá, cross the rapid stream, which here, as well as nearer the shore, before passing Plataniá, is shaded by lofty plane-trees, almost all with vines twining round them. I now begin an ascent which lasts for near an hour, when I arrive at the village of Thériso. The number of olive-trees about these villages of the Rhíza is not great; neither is there much arable land near them: the flocks of the district are however numerous. I pass through the straggling and well-shaded village of Thériso; and, continuing my course nearly due eastward, for upwards of an hour, arrive at Dhrakóna, one of the villages classed together under the common name of Kerámia.

It was in a village of this district of Kerámia, that blood was first shed in June 1821, when the Sfakians, and as many Rhizítes as possessed arms, raised the standard of the Cross. The súbashi of Kerámia is said to have been shot by the Christians, on the 17th (29th) of June. The following morning the Mohammedans issued out of Khaniá, one division of them going to Kerámia, and the other to Alikianó-Vatólako, a little below Láki. The Mohammedans carried off the head of a single Christian from the scene of their conflict at Kerámia, but had the worse of the engagement at both places. The Christians were always victorious, while the Khanióte-Mohammedans were without aid from the other cities. Hence the Austrian Consul, writing on the 15th of August 1821, says, "Les Turcs ont fait plusieures sorties et ont toujours été repoussés avec perte:" and thus the correligionaries of these unwarlike Moslems sang, as they approached Khaniá,

Ἐλᾶτ' ὑσεῖς οἱ Καστρινοί,
κ' ἡμεῖς οἱ Ῥιθυμνιῶταις,
νὰ πάμε, νὰ γλυτώσωμεν
τζὴ κλέφταις τζὴ Χανιώταις.

Press onward, all ye Kastrians,
With us Rhithymniótes,
That we may go and save from death
The cowardly Khaniótes.

Dhrakóna contains about thirty-eight houses: all its inhabitants are Christians, as is the case with nearly every village of the Rhíza. I remain here about half an hour. Soon after eleven o'clock we have before us Kendros, Mount Ida, the hills beyond the plain of Mylopótamo, and the bay of Rhíthymnos. Kendros now retains a mere sprinkling of snow, and the High Mountain himself, Pselorítes, as Ida is emphatically called by the modern Cretans, although for at least a thousand feet from his highest peak he had on a thick covering of snow only a fortnight ago, now shows strips of his rocky surface almost to his very summit. I pass Kámpi, the last village in the district of Khaniá, at half past eleven. Its Christian population is assembled at the church. Every one, of both sexes, is dressed in his or her gayest apparel. All are gladly bidding adieu to the last day of their Lent. I have seen many lambs this morning: some already slaughtered for to-morrow's feast: others tethered and destined to live only a few hours.

> Pleased to the last each crops his flowery food,
> And licks the hand just raised to shed his blood.

We are reminded of the Jewish paschal lamb, both by the Lampré or Easter-Sunday of the Greeks, and by the Kurbán-bairám or Feast of Sacrifice of the Mohammedans. Every true believer, whose means enable him to incur such an expense, is expected to kill a sheep on that occasion: its flesh, however, is not consumed in merry-making, but is, I am told, ordinarily distributed among the poor. As I noticed this morning the tethered lambs, on passing the cottages of Christians, so, for a few days before the Kurbán-bairám[1], I saw fine fat sheep tied up in half the shops of the bazár of Khaniá.

At a few minutes after one I arrive at Rhamné, and obtain some refreshment in the house of my former host[2].

[1] See above, p. 21.

[2] See Vol. I. p. 65.

At two o'clock I leave Rhamné, and, in about an hour and a quarter, arrive at Pemónia: in descending towards this village, I had a fine view of the plain of Apokórona, bounded by the Sfakian mountains on the right and by the gulf of Armyró before us, Mount Ida of course remaining in sight. Fré, one of the most considerable villages of Apokórona, is only about a mile distant from Pemónia: olive-trees cover the plain between the two places. At Fré I remained about an hour with the priest of the village: he was one of the deputies sent to Murniés last autumn, and also one of the delegates who carried petitions to the Consuls. It is said that his name was included in the list of the proscribed, but that remonstrances were made to the Pashá by some of those about his person, respecting the disgrace of suffering the execution of a man in whose house he had so often eaten bread and drank water. The Papás's house was for some time the Pashá's head-quarters, when the Arab troops were cantoned in the different villages of Apokórona. I leave Fré at half past five, and, after passing the village of Dzidzifé, I arrive at 'Ipos at a quarter before seven: I lodge in the house of a gossip of my Sfakian guide.

The corn-land in the plain of Apokórona usually gives a return of from fifteen to twenty fold the quantity of seed. I am assured that, not long ago, three measures of barley, produced a crop of nearly ninety measures. The olive-trees of Apokórona are generally small: they are very unlike the magnificent trees of Sélino: it is said, that for a thousand mistata no less than four thousand roots are here required. A wealthy Turk, who was sent into exile last winter, after the affair of Murniés, in which the government thought proper to suppose him implicated, possesses about sixteen thousand roots of olives, and their mean produce is not much above three thousand mistata. The villagers put the weekly consumption of oil, in each family, at four okes and a half. Each olive-tree sells here for from fifteen to twenty piastres: some may fetch thirty: a tree here and there, if unusually large and fruitful, may occasionally sell for more than double the latter sum; but of this last class there are few in Apokórona. If the arable field in which the olives stand is good, it will also be worth a hundred piastres the measure[3]. A calculation of the return made for the investment of capital at these prices, will give about thirty or even thirty-five per cent. as its annual amount: and this is not more than is really obtained. About three years' purchase is the ordinary price of land: I have met with cases where it has been obtained for a good deal less. The most ample security can be had for money, on mortgage of land and property far exceeding in value the sum borrowed: and, for a loan thus obtained, the rate of interest is sometimes as low as fifteen per cent.; in the case of a monastery it may be even less; but, for a private individual, it commonly amounts to nearer thirty. This is all quite natural; for the country is very fertile, and money, wherever it is applied to cultivate the soil and develope its dormant powers, produces so large a return, that the borrower can well afford to pay twenty and even thirty per cent. for the use of it.

Maniás and his gossip sit down this evening to a dish of snails[4], a luxury in which the Greek is allowed to indulge even on his most rigorous fasts. The snails of Crete are highly prized in the Levant; and they are one of the regular exports of the island. The Christian populations of Constantinople, Smyrna, and Alexandria all enjoy this delicacy during the fasts of the Oriental church.

'Ipo produces good wine, and I obtained some this evening: I also found wine last night at Láki: I did not once obtain it during the days which I have lately spent in the province of Sélino[5].

[3] Τὸ μισούρι.

[4] Κοχλιούς.

[5] See above, p. 147.

May 4.

This is the Easter-Sunday[6] of the Greeks: I arise, as usual, with the dawn, but find that Maniás and his gossip have been absent from the house near two hours: the ceremonies of the Greek church begin at as early an hour, especially on her great festivals, as did those of our own Roman Catholic ancestors in Old England.

> At midnight then, with carefull mind, they up to mattens ries,
> The clarke doth come, and, after him, the priest with staring eies[7].

The pious Greeks did not return till some time after sunrise, since they remained to assist at the ceremony of the resurrection[8], and at the burning of Judas. Each house of the village contributes a portion of the wood, of which a huge fire is made, and then, while the priest reads or chants the liturgy, a rude painting of Judas is thrown into the flames.

The modern Greek delights in representations of the suffering, burial and resurrection of Christ[9]. It was easier to transfer the ceremonies of paganism to a new object than wholly to abandon them. We have already seen, that the ancient Cretan used annually to commemorate the marriage of the two Supreme Deities

[6] Ἡ Λαμπρή.

[7] BARNABE GOOGE, The Popish Kingdom, fol. 52. quoted in BRAND'S Popular Antiquities, Vol. I. p. 138.

[8] Ἡ Ἀνάστασις.

[9] CHANDLER, Travels in Asia Minor, c. XLVII. "The Greeks now celebrated Easter. A small bier, prettily decked with orange and citron buds, jasmine, flowers, and boughs, was placed in the church, with a Christ crucified, rudely painted on the board, for the body. We saw it in the evening; and before day-break were suddenly awakened by the blaze and crackling of a large bonfire, with singing and shouting in honour of the resurrection." HUGHES, Travels in Greece and Albania, Vol. I. p. 405. "It was Easter Sunday—a solemn piece of mummery is this day peformed; when a wooden image, representing Christ's body, which had been buried in a kind of sepulchre, on the preceding Good Friday, with mournful lamentations, is now raised up and shewn by the papas to the people, who view it with every demonstration of joy, and salute it with long-continued exclamations of Χριστὸς ἀνέστη."

of Heaven, Zeus and Here, by a mimetic exhibition[10]. Thus the festival of Adonis was celebrated at Alexandria; and thus also at Cyprus a part of the religious ceremonies, by which Ariadne-Aphrodite was annually honoured, consisted in the imitation, by a young man reclining on a sofa, of a woman or goddess suffering the pangs of parturition[11].

As the Cretan Christian addressed his Mohammedan friend, a fortnight ago[12], with the words "Good Bairám, gossip[13]," so the Mohammedan now returns the compliment by a "Good Easter[14]," when he meets a Christian of his acquaintance. The ordinary Christian salutation, of the Oriental church, at this season, "Christ is risen[15]," with the uniform reply, "Truly he is risen[16]," cannot, of course, be exchanged between Christians and Mohammedans.

A little song, which I have heard sung by Maniás, mentions this "Christ is risen," as the peculiar characteristic of Easter-day, and is, on other accounts, interesting enough to deserve being recorded.

Each nightingale of Vlakiá,
  And, in the west, each bird,
At eve, and morn, and at mid-day,
  With plaintive note is heard,

Bewailing Adrianopolis,
  And her disastrous fate!
Since now no more the three great feasts
  She dares to celebrate.

[10] DIODORUS SICULUS, v. 72. and above, Vol. I. p. 204.

[11] PLUTARCH, Theseus, XX. p. 41. Reisk. Ἐν δὲ τῇ θυσίᾳ, τοῦ Γορπιαίου μηνὸς ἱσταμένου δευτέρᾳ, κατακλινόμενόν τινα τῶν νεανίσκων φθέγγεσθαι καὶ ποιεῖν ἅπερ ὠδίνουσαι γυναῖκες· καλεῖν δὲ τὸ ἄλσος Ἀμαθουσίους, ἐν ᾧ τὸν τάφον δεικνύουσιν, Ἀριάδνης Ἀφροδίτης.

[12] See above, p. 21. and Vol. I. p. 11.

[13] Καλὸ Μπαϊράμι κουμπάρη.

[14] Καλὴ Λαμπρή.

[15] Χριστὸς ἀνέστη.

[16] Ἀληθῶς ἀνέστη.

No more are tapers lighted up
On the birth-day of our Lord:
Nor, henceforth, will a single branch
The day of palms afford.

No more, on Easter-day, each voice
Of greeting in her streets,
"Our Christ is risen:" "True he's risen:"
Those blessed words, repeats[17].

It is evident that the event deplored in this ditty is the capture of Adrianople, by Amurath the First, and its conversion into the chief seat of his government and religion in Europe[18]. I suppose the Greek verses to be of nearly the same date as the event which gave rise to them: they have thus been orally preserved for between four and five hundred years.

Maniás and our host sit down to an early meal of roast lamb: an agreeable contrast to the privations of

17 Κλαίγουν τ' ἀηδόνια τῆς Βλακιάς,
καὶ τὰ πουλιά στὴν δύσιν,
κλαίγουν ἀργά, κλαίγουν ταχεά,
κλαίγουν τὸ μεσημέρι·
κλαίγουν τὴν Ἀδριανούπολιν,
τὴν βαρεά κρουσμένην,
ἀποῦ τὴν ἐκρουσέψανι
τζὴ τρεῖς ὀρταὶς τοῦ χρόνου.
τοῦ Χριστουγέννου γιὰ κηρί,
καὶ τοῦ βαΐου γιὰ βάϊα,
καὶ τῆς Λαμπρῆς τὴν κυριακήν
γιὰ τὸ, "Χριστὸς ἀνέστη."

18 GIBBON, Decline and Fall, c. LXIV. HAMMER, Geschichte des Osmanischen Reiches, Vol. I. Book V. p. 164. (ed. Pest, 1827.) gives the year 763 of the Mohammedan era, or 1361 of our own, as the date of the capture: in this date the great majority of the oriental authorities are agreed.

their long fast. At Easter[19], and on the arrival of other festivals which succeed to fasts, "the Greeks fail not to make compensation for extreme temperance by the opposite license. Thus their life is passed in an alternation of extravagancies, and the priest who enforces the one excess does little to repress the other[20]."

The sun shines brightly this morning, although he gives none of those "signs of gratulation" which an old English superstition attributes to him on this day. Sir Thomas Browne writes very gravely[21], "We shall not, I hope, disparage the resurrection of our Redeemer, if we say the sun doth not dance on Easter-day. And though we would willingly assent to any sympathetical exultation[22], yet cannot conceive therein any more than a tropical expression."

[19] LEAKE, Travels in the Moréa, Vol. III. p. 209. "At Patra I saw the boatmen, on Easter Sunday, ranged in parties along the beach; each boat's crew seated on the ground in the hot sun, round a great fire, roasting lambs, and waiting with impatience after the forty days' fast, till they were dressed. By nine or ten o'clock in the forenoon, most of the families in the town had dined, and were already half drunk and dancing."

[20] WADDINGTON, Present Condition and Prospects of the Greek Church, p. 57. RICAUT, The present State of the Greek Church, p. 136. "The severity of their Lents is more easily supported by the expected enjoyment of the following festival; at which time they run into such excesses of mirth and riot, agreeable to the light and vain humour of the people, that they seem to be revenged of their late sobriety, and to make compensation to the Devil for their late temperance and mortification towards God."

[21] Sir T. BROWNE, Works, p. 221. ed. 1686. Vulgar Errors, Book v. §. 17. The notion is alluded to in an old ballad quoted by BRAND, Popular Antiquities, Vol. I. p. 137.

But, Dick, she dances such a way,
No sun upon an Easter-day
Is half so fine a sight.

On the next page, of the same work, is quoted, from the "British Apollo,"

Old wives, Phoebus, say,
That on Easter-day
To the music o' th' spheres you do caper:
If the fact, sir, be true
Pray let's the cause know,
When you have any room in your paper.

[22] Thus SOPHOCLES calls Dionysos ἄστρων χοραγόν, and EURIPIDES says, that, during the celebration of an ancient religious festival,

ἀνεχόρευσεν αἰθήρ,
χορεύει δὲ σελάνα·

see LOBECK, Aglaophamus, Vol. I. p. 218.

Here, however, we are celebrating the festival of Easter more than a month after its anniversary in all Roman Catholic and Protestant countries, so that, even if we believed, with our superstitious countrymen of former days, that the sun dances for us in March or April, still we could hardly expect him to do the same thing again, in the month of May, for the Oriental church.

Miraculous phenomena, supposed to have taken place on the anniversary of Easter, have been referred to in the great controversy on the proper time of celebrating that festival. Had the English superstition, respecting the sun's dancing on Easter-day, prevailed in this part of the world, it might have served as an additional argument. The oriental Christian appealed to the annual descent from heaven of the sacred flame, which served to illumine the holy sepulchre, on his Easter-Sunday, and thought this occurrence an unanswerable argument against the practice of the western heretics.

Phenomena, still more extraordinary than that of the holy flame, were also said to happen in Egypt, during the forty days which follow the paschal festival of the eastern church. Graves opened and bodies of dead men arose, during the whole of that period. When once the orientals adopted the observance of the western church, the graves no longer opened in Egypt, and the sacred flame no longer shewed itself in Jerusalem: but, on their return to their own ancient usage, "the holy flame again descended, and the earth again gave up its dead[23]," yearly, as it had been in the habit of doing previously[24].

[23] This feature of the legend was manifestly suggested by the narrative of MATTHEW, XXVII. 52.

[24] CHRISTOPHORUS ANGELUS, de vit. et mor. recent. Graec. c. XLII. (in LEO ALLATIUS, de quorund. Graecor. opinat. p. 181.) "Αρχεται ἡ γῆ ἐκείνη ἀναπέμπειν τὰ νεκρὰ σώματα, ἀπὸ τῆς μεγάλης πέμπτης, τουτέστιν ἀπὸ τῆς ἡμέρας ἐν ῇ ὁ Χριστὸς ἐποίησε τὸ δεῖπνον τὸ μυστικόν, καὶ καθ' ἑκάστην ἡμέραν ἐκβάλλει τὰ νεκρὰ σώματα, ἕως τῆς ἡμέρας τῆς ἀναλήψεως—τοῦτο δὲ τὸ θαῦμα Τοῦρκοι καὶ "Ελληνες διηγοῦνται, οἵτινες πρὸς

ἐκείνους

I leave 'Ipo at a few minutes before nine: in three quarters of an hour I cross the river which flows from the Sfakian mountains and falls into the sea below Armyró: its stream is rapid, and its water is here deliciously cold. At ten o'clock we arrive at the village of Prósnero, the last before the Sfakian frontier.

At Prósnero is seen the ruined pýrgo, or tower of Alidhákes, a Mohammedan who defended himself for some time, from within his little castle, against the Christian insurgents, at the outbreaking of the Greek revolution. When the Sfakians first arose, a division of about two hundred and fifty of them marched down to attack this tower. The father of Alidhákes had been killed by the Sfakians in the war which took place in Crete seventy years ago, (after the attempt of Russia on the Moréa,) and Alidhákes was now prepared to make a vigorous resistance, having with him, within the tower, about eighty picked men. Although the Sfakians transported some old cannon, which they found in the fort at Armyró, to this spot, they did not succeed in carrying the pýrgo; and, after losing some time in the attempt, at last left a smaller party to blockade it. When afterwards abandoned, it was dismantled by the Christian insurgents.

The ascent becomes steeper after leaving Prósnero: at half-past ten I stop and repose for twenty minutes under the shade of a tree. At a quarter past eleven we reach a small plain, called Krápi, where the Rhíza or Rhízoma ends: here therefore we enter the higher ranges which are called Sfakiá. There are many ilexes[25] and some wild-pear-trees[26] on this little plain of Krápi.

On leaving the plain we begin to enter the gorge or pass which is to conduct us to Askýfo. The moun-

ἐκείνους τοὺς τόπους πορεύονται. ὡσαῦτως καὶ οἱ προσκυνηταὶ πορεύονται ἀπὸ Ἱερουσαλὴμ ἕως ἐκείνου τοῦ τόπου ἰδεῖν τὸ θαῦμα——ἐδοκίμασαν οἱ Ἕλληνες ποτὲ ἀπὸ τοῦ νῦν εἴκοσι ἔτη πρότερον ποιῆσαι τὸ πάσχα κατὰ τὸν νέον ἀριθμόν, ἀλλ' οὐδαμῶς ἐξήρχετο τὰ νεκρὰ κατὰ τὸν νέον ἀριθμόν, &c. &c. &c.

[25] Πρῖνοι.

[26] Ἀχλάδαις.

tains on either side are lofty, and generally present a stony and barren appearance: ilexes, however, are spread over the greater part of them.

As we passed along, Maniás described to me, at great length, the disastrous flight of the Mohammedans through this pass, at the very outbreaking of the revolution: he even pointed out spots where individuals, of whom he spoke, had fallen: and the whitening human bones, which were still seen in some of the places, were a painful proof of the faithfulness of his story. I heard the same facts described by many other persons before I left the island, and the following is the sum and substance of what I learnt.

The Pashá determined, in August 1821, on penetrating into Sfakiá, and accompanied by an immense host of Cretan Mohammedans, made good his passage to Askýfo. All attempts to withstand the advance of the enemy were vain, and the Sfakians were obliged to retire to the loftier mountains to the west.

The Mohammedans encamped in the plain of Askýfo, expecting to receive, ere long, the submission of the rebels. The Greeks, however, in no wise dispirited by the overwhelming numbers of the enemy, employed diligently that evening, and the ensuing night, in summoning aid from all the neighbouring villages, and, most of all, in sending intelligence of what had happened to Maláxa, an important post, where nearly a hundred Askyfiótes were stationed, along with other Greeks. Thus, by daybreak, the Christian force assembled at Xerókampos[27], about two miles west of Askýfo, was between four and five hundred strong.

On advancing towards the plain, they saw that it was covered by the Turkish tents and troops, which, since a breathless calm prevailed, were more than half concealed from their view by the smoke of their villages, which were already sacked and in flames. Rússos, ac-

27 Ξηρόκαμπος.

companied by a small party, approached his own house, the highest on the side of the mountain, though little more than a musket-shot from the edge of the plain: on entering it he found that a large earthen vessel of wine had escaped the notice of the Mohammedans. This was soon distributed among his followers, who, descending a little nearer the plain, opened their fire from behind some low walls; and many other Sfakians, who had already occupied the villages of Pétres, and Stavrórakhi, situated a little to the south of the station of Rússos, and on the same western slope, began their attack at the same time. These simultaneous volleys caused no little astonishment and rage among the Mohammedans, who, on seeing the Greeks approach, supposed them to be coming to submit[28], and that they should have no more trouble in putting down what they called the Sfakian revolt.

The action thus commenced soon after sunrise: for full seven hours the Mohammedans constantly kept up a heavy fire of musquetry, and directed, as well as they were able, against the redoubts[29] of the Greeks, three field-pieces, which they had brought with them. During this time also considerable execution was done in the Turkish ranks by the less noisy fire which was directed against them, with a far surer aim, by the Greeks. By about two hours after mid-day, the Sfakians, were nearly nine hundred strong, numerous reinforcements of small parties having arrived during the morning. A considerable body of the mountaineers now took possession of the wood immediately above Karés[30], at the north-western extremity of the plain, and opened a destructive fire on the Turkish column which was already engaged at that village. The Mohammedans, engaged there, on finding themselves exposed to this unexpected attack, wavered for a moment, and then commenced a precipitate retreat. The flight of this division immediately caused a panic

[28] νὰ προσκυνήσωσιν. [29] τὰ ταμπούρια. [30] Καραίς.

throughout their whole host, and the rout became general. The Greeks pursued their enemies across the plain, and harassed their rear as they retreated along the path which conducts to the district of Apokórona.

As soon as the Mohammedans reached the very commencement of the descent, where the glen begins to contract in width, the Greeks detached from their main body a party, who, making a rapid circuit over the hills on the north-west, arrived on the left flank of their fleeing foes at the narrowest part of the gorge, about three hundred paces from its opening, and when the foremost of the fugitives were only just passing the spot. This party of Sfakians immediately opened, on the dense retreating column, a deadly fire, of which every shot told. The Mohammedans, with more courage than prudence, attempted for a moment to make a stand; but soon found, that the fearful advantages possessed by an enemy, who was almost entirely concealed from their view, left them no chance of safety, except in a rapid flight. Many of the horsemen at once abandoned their steeds, which, when they had to traverse the rugged paths of Sfakiá, only impeded their escape, and betook themselves to the mountains on the eastern side of the defile.

Before the last of the fugitives emerged from the glen, heaps of dead bodies lay on the road all the way from this narrow pass to the entrance above Krápi: and those who have experienced the intolerable stench produced, under the burning summer-sun of this latitude, by the decomposition of a single dead body, lying exposed to its putrefying beams, will readily believe the Sfakian who says that for weeks afterwards they were unable to pass even near the spot.

> Tal puzzo n'usciva
> Qual suole uscir dalle marcite membre.

The traveller who visits Sfakiá, as I do, thirteen years after the event, still sees by the roadside the bleached bones of many of those who fell: memorials of mortal strife, which, in these regions, are but too frequently met with.

When the main body of the Mohammedans had at length passed out, and found themselves on the open barren mountains about Krápi, the rear continued to be harassed by a steady pursuit, to avoid which numerous parties made their escape to the mountains. The Sfakians pursued the flying enemy as far as Armyró, near twelve miles from Askýfo, harassing his rear, and cutting off every one who was not active enough to keep up with the main body in their disorderly flight. At Armyró sunset and the shades of night gave the Turks that protection which they might have found in their own numbers, had their leaders been but once able to rally them when they had reached the plain.

But, though the main body of the Turks was thus unmolested by any further pursuit, their numerous stragglers had other and greater horrors to endure. Those of them who were wounded were unable long to continue their flight; and, with no friend to aid them, sunk down and died. Even the strongest and most vigorous were scarcely better able to escape; for most of them lost their way in the mountains, or were intercepted by parties of Sfakians, who, for the next two or three days, according to their own accounts, hunted these unhappy stragglers "like so many wild-goats." Corpses not stript of their arms, when occasionally found by Sfakian shepherds for months afterwards, sufficiently indicated that others had died from absolute want and exhaustion. Some, after skulking under the trees in the day-time, and vainly endeavouring to find their way towards Rhíthymnos by night, unable to withstand the stern behests of those imperious tyrants, Hunger and Thirst, at length, as their only chance of preserving life, entered a village, and, throwing themselves at the feet of the first Greek they found, implored mercy and a draught of water. Perhaps the reader may suppose that such a prayer would be granted: if he does, he knows not the Sfakian, whose hatred of the very name of Mohammedan makes him even proud of recounting such atrocities as these massacres of unresisting and helpless suppliants[31].

A word more, and the history of this disastrous flight will be concluded. As if it had been destined, that all modes of escape should prove equally unavailing, many parties of Mohammedans, who fled in bodies of from five or six to eighteen or twenty men, were intercepted by the mountaineers: one was thus destroyed at Límni, and another at Gonía, between Kalikráti and Pólis: a third, more numerous than the other two, had got right across the mountains to Skaloté, when at length they were observed, were attacked by a superior force of the active and warlike mountaineers, and were cut off to a man.

The Mohammedans are said to have lost, on the occasion, nearly nine hundred men, as well as many mules laden with military stores, and their three field-pieces. The loss of the Christians was quite insignificant.

It may easily be imagined how so signal a victory would elate the spirits of those who had already a strong religious feeling on the subject of the war: who now believed themselves to be under the especial guardianship of the Deity, to whom they wholly ascribed their success: who were fighting "for the body and blood

[31] A solitary Mohammedan entered a Sfakian village at mid-day: all its male inhabitants happened to be absent: the women (who were unarmed, while he had arms) at first treated him kindly; but only till they found an opportunity of dispatching him. When I visited Nípro, my host said, that, two or three days after their great victory, "a Mohammedan came here, and fell on his knees a few paces from my door, imploring a draught of water." And what did you do? "I took my tufék—and shot him." We must make some allowance, in estimating the morality of such conduct, for the provocation received by the Christians of Crete, during more than a whole century; and, most of all, for the feelings so recently excited by the butchery of the Patriarch at Constantinople; of the Bishop of Kísamos at Khaniá; of the Metropolitan and five other Christian Prelates at Megálo-Kástron, and of nearly *a thousand* unarmed Greeks in various parts of this island. See, on this subject, the Additional Note at the end of the Chapter.

of the Lord Christ:" whose every banner was accompanied by a priest, often one of the bravest combatants: who celebrated, with an unheard of frequency, the most solemn ordinances of their religion: and who were thus raised to so exalted a pitch of religious fervour and enthusiasm, that, in their zeal for God's honour and service, they could, one and all, subdue even their dearest affections, when they believed that, by indulging them, they would be rendered unworthy champions of the holy cause which they had espoused[32].

An ascent of about forty minutes from Krápi, brings us to the highest point of the path, and, in a few minutes more, we see the plain of Askýfo, spread out before and below us. We are rather more than a quarter of an hour in descending to it. The villages classed together under the common name of Askýfo, are situated in a plain of about three to four miles in circumference, so surrounded by lofty mountain-summits, that it has somewhat the appearance of a large amphitheatre[33].

Maniás being still unmarried, his mother keeps house for him: we find, however, that the door of his dwelling at Askýfo is closed, the old lady not having yet come up into Sfakiá, from their winter residence at Dhrámia[34]. I therefore lodge with a neighbour and nephew of Maniás. My host's wife is extremely beautiful: she was born at Anópolis, and never left her native place until she married, three years ago: since that event she has not stirred out of Askýfo, except once to visit her mother at Anópolis.

The difference in temperature between the plain of Apokórona this morning at ten, and these lofty summits

[32] The Christian husbands of Crete, on thus becoming soldiers of the Cross, shrunk from the caresses of their wives, as from a pollution, which would most probably be punished by their falling in the next engagement. This singular piece of religious self-denial lasted, with most of them, for the greater part of the first year. I believe this feature is one of several peculiar to the war *in Crete*. The religious principle was certainly stronger here than elsewhere: see above, Vol. I. pp. 66–67.

[33] See the engraving below, at p. 188. [34] See above, Vol. I. p. 73.

even at two in the afternoon, is very considerable. There is always an agreeable breeze here in the hot weather, and the evenings and mornings are delightfully cool, even in the months of July and August.

I suppose Askýfo to be more than fifteen hundred, perhaps nearly two thousand feet, below the highest summits of the Sfakian chain of mountains, and to be somewhere between four and five thousand feet above the sea-level.

The different hamlets spread round the edges of this plain, and denoted by the common name of Askýfo, are called respectively Goní, where I have taken up my quarters, Péra-Goní, Pétres, Mudhári, Kóstos, Stavró-rakhi, and Karés.

The present number of families at Askýfo is about one hundred and sixty. They are Christians, without a single exception, as is the case throughout Sfakiá.

The town of Sfakiá[35] is the chief winter residence of the Askyfiótes: many families, however, not having any dwelling except at Askýfo, make the necessary provision for remaining here through the winter, and are ordinarily confined to their houses, for several weeks, by the snow[36].

Soon after arriving at my present quarters I asked for a glass of water. My host, instead of saying "Your health[37]," to me, in the usual way, after I had swallowed it, exclaimed, "Christ is risen," to which I, of course, replied, "Truly he is risen." This Easter salutation is now *alone* used. Even when a man sneezes, "Christ is risen," are the words by which he is addressed.

Several of the villagers come to my host's house to see and talk with me. One of them is an uncle of Maniás, and brother of Búzo-Márko, who fell at Grabúsa. Of the seven Sfakians who were executed by the Pashás last December, four were of Askýfo, and

35 'Η χώρα τῶν Σφακιῶν.

36 See Vol. I. p. 73.

37 Μὲ ὑγεία σας.

only one of them had been present in the assembly at Murniés. The other three of the seven were of Nípros, Komitádhes and Hághios Ioánnes: two of the Askyfiótes, and the other three Sfakians, were arrested and executed in this village in the following manner.

When the Pashás had established their head-quarters at Fré, and their army was dispersed through the villages of 'Ipos, Prósnero, Alíkampo and other places in the plain of Apokórona, they at length sent to Askýfo Alí-efféndi, the Turkish President of the Council of Sfakiá, accompanied by a bímbashi, (colonel), with a battalion of regular troops and about one hundred and fifty Arnauts. On their arrival the President and the Colonel professed to be the bearers of letters[38], both from the Pashás in the plain below, and from the Viceroy, Mehmét-Alí-pashá, to the inhabitants of Sfakiá. They therefore despatched messengers to each of the villages of the whole Sfakian district, requesting that two or three of the principal persons of every place would come to Askýfo, in order to be present at the reading of these letters. The wary Sfakians suspected some treachery, and came but slowly; so slowly that the Turks waited no less than six days before they executed their real intention. They never read any letters from the Viceroy or Pashás, but, on the evening of the sixth or seventh day of their stay, about two hours before sunset, they arrested five of their destined victims, and immediately wrote to the Pashá, telling him the names of those whom they had secured.

The sensation produced in the village may be easily conceived. These hardy mountaineers, who had maintained a contest with the Turks, with but little intermission, for a period of ten years; and who, however closely pressed by superiority of numbers, had always found a refuge in the fastnesses or on the summits of their native mountains, were not likely to stand tamely

[38] Γράμματα.

by, and see their innocent fellow-villagers and relations arrested and conducted down to the head-quarters at Fré, probably to be executed in cold blood by these Egyptians, whom every one hated as bitterly as the old Turks.

Now the villagers of Askýfo and Nípros, not an hour distant, would supply a body of full two hundred men: they would also be able to obtain a reinforcement from Asfénto, so that probably they would muster about three hundred, nearly all of whom could arm themselves in an instant[39]. The leading men among them held a secret deliberation, and felt satisfied that daybreak would see their fellow-villagers marched off under a strong escort, of perhaps a hundred men, to the Pashás. They therefore determined to go in two parties so as suddenly to inclose them as they passed the narrowest part of the defile. "We will then demand their prisoners, and we will have them either dead or alive[40]." Happily, however, for the tranquillity of Crete, and, perhaps, for the Sfakians themselves, the messenger sent to the Pashás returned in the silence and darkness of night, and communicated his orders to the President and Colonel. When the Sfakians were on the look out, at the very earliest dawn, to see what the Turks were on the point of doing, the first sight that met their eyes was the five bodies of the unfortunate prisoners, one or two of whom had been despatched by the bowstring, and the rest had been bayonneted by the Arabs, in the house where they were guarded.

The Askyfiótes around me, three of whom, in their height and muscular strength, are such men as one sees only in these mountains, sigh as they repeat the tale, and point out to me, in the group which surrounds me,

[39] Although the whole of Crete was nominally disarmed, when the Egyptians had the country given to them by England, France, and Russia, yet the Sfakians nearly all of them still have concealed arms, and could at any time rise equipped for their old pursuits of war and plunder.

[40] Ἢ ἀποθαμένους ἢ ζωντανούς.

a young woman, drest in the garb of mourning, who lost her father during the war, and whose brother was one of the victims thus recently sacrificed. His orphan, a child of four years old, is standing by her. The Sfakians doubt whether they did not do wrong, in thus remaining quiet, and think that their best course[41] would have been to have taken their arms, and once more occupying the most inaccessible of their mountains, thence to have waged and suffered a war of extermination with their foe.

I inquire whether any Askyfióte remembers the events of the Russian invasion of the Moréa in 1769, when nearly a thousand Sfakians left their homes to fight against the Turks, and when the Cretan Turks, taking advantage of their absence, exacted a heavy penalty from those whom they had left behind them, for the part they took. I am told that, among the five men seized and massacred here, there was an old man, Andhrúlios Papadhákes, who was one of Lambro's Capitanéi, and still possessed, at the time of his death, Lambro's pistols and some of his clothes. He was mentioned in the Greek song which celebrated Lambro's exploits[42]. The poor old man, whose extreme age, for he had lived nearly a whole century, might alone have protected him, even if he had taken some part in an assembly where thousands met together, had never been present at Murniés. But he was distinguished from olden times, and even among the inhabitants of these mountains, for his hatred of the Turkish rulers of his country ; and the reputation which he thus acquired was probably the cause of his death, and certainly entitles him to be regarded as one of the patriots, who were celebrated by our poet before the outbreaking of the Greek revolution.

[41] Τὸ *καλλιώτερον*: the common modern Greek form is *καλλίτερον*: examples of both forms are given in Du Cange: on *καλλιώτερον* see Phrynichus, p. 136. and Lobeck.

[42] Τὸ *ὁποῖον εἴχασι βγαλμένο διὰ τὸν* 'Ραμπρό.

And some to higher thoughts aspire,
  The last of Lambro's patriots there
  Anticipated freedom share;
And oft around the cavern fire
On visionary schemes debate
To snatch the Rayahs from their fate.

The schemes of the Cretan patriots have indeed proved visionary! notwithstanding the sacrifices they made, and the heroism they shewed in their long and obstinate contest.

To complete the tale of the seven Sfakians executed, two remain to be spoken of. They were seized and put to death at Rhíthymnos, where they were busied in traffic: one of them had just returned from Constantinople. The cruelty and treachery of the Egyptian authorities in thus punishing the innocent in order to strike terror into all, need no comment.

ADDITIONAL NOTE

TO

CHAPTER XXXIV.

THE following extracts, from the Archives of the French Consulate, at Khaniá, will alone exhibit a tolerably clear outline of the condition in which the Christians of Crete found themselves for many years before the outbreaking of the Greek revolution.

I will give a single passage, of ancient date, to shew how soon the Khanióte Mohammedans acquired the character which they so long maintained. In the year 1726, a Frenchman was assassinated in the town of Khaniá. The French Consul says, at a "délibération," with "the nation," as the resident French merchants call themselves, "Nous connaissons la mechanceté des habitans de cette ville, naturellement portés à la rebellion, et appuyant toujours leurs compatriotes, à tort ou à raison, *menaçant s'il arrive le moindre mal à l'auteur du meurtre, de mettre en pièces tous les français: ceux mêmes, avec les quels nous faisons du negoce, n'en parlent pas autrement.*"

I will now pass on to the few years preceding the Greek revolution. "*Le* 30 *Avril*, 1816. Les peuples de ces contrées ne ressemblent en rien aux autres peuplades de la Turquie. Le Turc habitué, envers le Grec, à un génre de vie despotique, y est vain, insubordonné, inclin à la révolte," &c.

"*Le* 15 *Juin*, 1819. Les 8, 9, et 10 du courant, des Turcs excitèrent une insurrection, en parcourant les villages circonvoisins, et en forçant les habitans des campagnes de se joindre à eux pour marcher sur la Canée.

"Le 11 une vingtaine d'hommes s'introduisirent dans la ville. Le 12 au matin, 400 des mêmes y entrèrent par bandes, et se portèrent sur les remparts, où ils convoquèrent une assemblée, dans un des kiosques qui s'y trouvent, et forçant toutes les classes de la population à s'y rendre, et notamment les principaux du pays.

"Cette assemblée commença ses délibérations par prononcer la destitution du Pacha, et procéder à la nomination d'un nouveau Musselim pour gouverner la ville dans la personne du Janissaire-Aga : elle prononça la peine de mort contre les grands du pays qui iraient voir le Pacha, ou qui auraient la moindre intelligence avec lui.

"En rapprochant toutes les circonstances des données que j'ai procurées, il parait que le principal motif de cette révolte vient de ce que le Pacha, après avoir contraint, dit-on injustément, les Grecs et les Juifs—à donner des huiles pour Constantinople, voulût y contraindre également les Agas, pour compléter un chargement. Ceux-ci, *habitués à faire supporter toutes les charges aux Rayas*, ont provoqué la révolte, dans l'espoir d'intimider le Pacha."

"*Le* 28 *Août*, 1819. Le 24 courant Ibraim Pacha arriva en ce port pour lui succéder. La presence du gouverneur, et la réputation qu'on lui donne en assurant qu'il a fait étrangler beaucoup de monde, pendant qu'il occupait la place de Janissaire-Aga à Constantinople, *ont frappé de terreur les grands assassins de cette province*, et leur ont fait prendre la fuite."

"*Le* 18 *Decembre*, 1819. La soldatesque de cette ville vient de se permettre comme à son ordinaire, un acte d'insubordination des plus marquans, en refusant de se soumettre aux volontés du Grand Seigneur, contenues dans deux firmans emanés du Gouvernement de Sa Hautesse.

"Ces firmans portaient en substance, la peine d'exil des 4e et 22e corps de Janissaires en garnison à la Canée." (Here follow details of their refusal to embark.)

"Dès lors, (l'arrivée du nouveau Pacha,) les mauvais sujets qui font tous partie des corps de Janissaires, et à qui on donne le surnom de Palikári (braves) parcequ'ils ont été des assassins, ont craint, en se soumettant à l'exil, de se voir forcés de suivre la marmite, ou de s'exposer, en restant, à être sans appui et à la merci du Pacha, qui leur ferait un mauvais parti. Les Agas de leur côté voyaient avec inquiétude le départ des satellites dont ils se servent pour intimider au besoin les Pachas, et pour maintenir leur morgue envers le reste de la population. Ce qu'il y a de certain c'est que, *si de tems à autre le gouverneur turc ne fait pas tomber quelques têtes en cette ville, on n'y jouira jamais de la tranquillité.*

"*Le* 15 *Mars*, 1820. Le Pacha fatigué de la resistance, que lui opposent les autorités locales, paraît se relâcher de ses prétentions, pour se livrer lui-même à des avanies contre les rayas. Ce changement de conduite du Pacha a enhardi les mauvais sujets, et les assassinats ont recommencé."

"*Le* 22 *Novembre*, 1820. En rendant compte à V. E. du résultat des instances que j'ai faites auprès de l'Autorité locale pour obtenir justice et réparation d'un pareil attentat, je crois devoir lui faire connaître le genre de gouvernement de ce pays, et l'influence plus ou moins forte de chaque autorité qui le composent, afin de lui donner une idée de l'impossibilité qu'il y a d'obtenir en ce pays des satisfactions suffisantes pour en imposer aux malveillants, et faire cesser cet état de crainte, où l'on est continuellement, d'être assailli et assassiné.

"Pendant l'intervalle de tems que les Pachas mettent pour se remplacer, la ville est gouvernée par un Musselim choisi parmi les Agas en charge, qui, pour ne point être contrarié dans l'exercice de ses fonctions, et pour maintenir l'esprit de corps avec ses collégues et principalement avec les chefs des régimens de Janissaires, se laisse influencer par eux.

"Les Agas possèdent toutes les terres seigneuriales du Grand Seigneur, moyennant une rétribution à vie, qu'ils renouvellent de père en fils, et sont par conséquent les plus riches propriétaires du pays, et les seuls qui approchent le plus des Pachas.

"Le corps des Janissaires est composé d'hommes du pays, de tout état, et sans solde, les quels sont en général méchans et cruels. Ceux qui ont commis le plus d'assassinats sont recherchés par les régimens, et jouissent de la protection entière de leurs Chefs, et des Agas, qui s'en servent au besoin, soit pour assommer à coups de bâton, ou faire assassiner ceux qui leur déplaisent, soit pour susciter des révoltes contre les officiers superieurs de la Porte, tels que Pachas, Janissaire-Agas, Mufti et Cadi, qu'ils suspendent de leurs fonctions, ou embarquent ignominieusement.

"Ces violences et la crainte d'être privé, par l'astuce et l'impudence des Agas, et des Chefs de régimens, des avantages attachés aux charges qui s'achètent en Turquie, obligent les Janissaire-Agas, Muftis, Cadis et autres officiers de la Porte d'avoir beaucoup de ménagemens pour eux, et de céder à leur volonté: *de manière que l'autorité du Pacha est, pour ainsi dire, nulle en ce pays, les Agas et les Chefs de régimens font les avanies qu'ils veulent sans que personne n'ose s'en plaindre: les malfaiteurs de chaque corps se livrent à toute sorte d'excès, et se permettent même d'entrer d'autorité dans les maisons des Rayas, et d'en disposer entièrement à leur grè*[1].

[1] The *details*, exemplifying the concluding assertion of this sentence, with which I became acquainted during my stay in Crete, would fill a volume. Many of them, however, are such

"Avant l'arrivée d'Osman Pacha ils se sont même permis d'envoyer des balles de pistolet pliées dans des morceaux de papier portant indication d'une somme à payer, *et si ceux à qui ces sortes de billets étaient adressés ne les acquittaient pas, ils étaient incontinent assassinés*[2]. Ces hommes pervers ont été jusqu'à se faire un jeu de tirer des coups de fusil, chargé à balle, de dessus les remparts de la porte de la ville, sur les individus qui s'y présentaient pour y entrer, *et de parier entr'eux que la victime tomberait de tel ou tel côté*[3]."

At Megálo-Kástron the position of the Christians was much the same as in Khaniá. The following are extracts from the correspondence of M. Boze, the French Consular Agent at the former city, with M. Roussel, then French Consul at Khaniá.

"*Le* 4 *Août*, 1806. Il regne ici un désordre redoutable. Les scélérats commettent toutes sortes de crimes. On n'entend parler que de massacres et d'injustices."

"*Le* 27 *Mars*, 1807. Aujourd'hui des scélérats Turcs ont encore assassiné, à coup de sabre, un bel homme Grec, dans le temps qu'il labourait à sa vigne à Stavrakia, à deux lieues d'ici. Ce meurtre a encore resté impuni comme tous les autres."

"*Le* 30 *Mars*, 1807. Les scélérats viennent encore de tuer deux pauvres Grecs à un village près de la ville. Le Pacha a envoyé un bairak pour démolir les maisons des assassins, et confisquer leurs biens."

"*Le* 18 *Juin*, 1807. On ne cesse ici, de même qu'aux villages, d'estropier, blesser et assassiner."

"*Le* 9 *Fevrier*, 1812. Assan Effendi, Tefterdar, a confié dernièrement à mon fils, qu'il avait expédié un de ses gens à Constantinople, pour y porter ses plaintes contre les Caniotes Turcs, au sujet *des grands massacres qu'ils font journellement aux pauvres Rayas*, qui se sont tous dispersés."

The French Vice-Consul writes as follows, at Megálo Kástron, on the 16th October, 1817. "Une incendie éclata avant hier sur les trois heures

such obscenities and horrors, as I could not attempt to describe: such fearful combinations of lust and cruelty, as one could hardly have believed human nature, in its most savage and degraded state, capable even of imagining.

[2] This was a *very common* occurrence: there were few parts of the island in which I did not hear of it.

[3] A wager, not unfrequently made by these *Mohammedans*, on taking this diversion at the expense of any ill-fated Christian who might happen to be passing, was an oke of *wine!*

après-midi, par l'explosion de quelques barils de poudre emmagazinés chez des Grecs. Plusieurs Turcs ont été victimes de l'explosion; et d'autres, au premier moment, *voulaient immoler tous les Grecs des environs.* Le Pacha fit mettre ces Grecs en prison, pour les sauver.

"Hier soir un Turc, qui en avait tué un autre, devait subir la peine de son crime. Le peuple s'amassa, se porta à la prison du Grand Château, et le délivra. Fier de son succès, il demanda que les Grecs, prisonniers pour le fait des poudres, leur fussent livrés pour être pendus, et immolés aux manes des Turcs, que l'explosion avait tués. *Le Pacha resista:* le palais (that is the serai of the Pashá) fut investi: le peuple repoussé: à neuf heures, tout semblait appaisé. A dix heures le tumulte recommença plus fort qu'auparavant, et dura jusqu'à minuit. *Le Pacha convint qu'il leur livrerait les Grecs à l'aube du jour.* A trois heures du matin on se porta de nouveau au palais: *les prisons furent forcées, trois Grecs en furent retirés, et pendus à leurs boutiques.*

"Depuis lors on dit que les mutins demandent encore 4 ou 5 Grecs: que de son côté le Pacha redemande le meurtrier que les mutins ont délivré hier soir. A défaut de le satisfaire il se prépare à partir."

A few extracts from the Correspondence of the Austrian Consul with the Internuncio at Constantinople, will partly shew what had been done by the Turkish authorities, and by the Mohammedan populace of Khaniá, during the months of May, June, and July, 1821:

"*A La Canée, le* 14 *Mai*, 1821. Les Turcs ont enfermé l'Evêque de Kissamos, le rendant garant de ce que pourroient faire les Grecs de son arrondissement." "*Le* 4 *Juin*, 1821. L'Evêque de Kissamos—a été livré à la fureur du peuple—qui sans égard pour son caractère l'a trainé dans toute la ville, à demi-nud, par la barbe, et l'a cruellement pendu sur le chemin de son arrondissement. Je ne saurais dépeindre la joie feroce du peuple, et encore moins le dépit des bourreaux, qui, malgré leurs cruautés n'ont pu faire pâlir le malheureux Evêque, digne d'un meilleur sort. Les Turcs ont demandé au Pacha la permission de massacrer impitoyablement les Grecs."

"*A La Canée le* 6 *Juillet*, 1821. Le peuple de la Canée a enfin obtenu du Pacha la malheureuse autorisation de massacrer tous les Grecs: on a même lû dans les mosquées un Fefta pour enhardir les moins feroces. C'est le 30 Juin, dernier jour du Ramazan, que le peuple a assouvi sa rage sur les malheureux Grecs restés dans la ville sous la protection de divers Agas leurs maîtres. Le nombre des victimes est evalué à trente environ."

"*A la Canée, le* 15 *Août*, 1821. Dans la nuit du 30 Juin au 1er Juillet la populace est sortie de la ville, poursuivant toujours les Grecs; on les noyait, on les étranglait, on les brûlait, révoltés ou non, tumultueux ou paisibles. On courût partout sur eux, comme sur des monstres dont il fallait purger l'île. On incendia une vingtaine de villages. Chaque jour on voyait au marché (et ceci dure encore), des jeunes femmes, qui étaient redevables de leur vie à leur beauté, mise à prix par leurs ravisseurs. Ce n'est que quelques jours après, que les Sfakiotes sont accourus, eux seuls qui avaient refusé de livrer leurs armes. Ils ont de suite établi un cordon de troupes sur les montagnes qui dominent la ville, se sont emparés des gorges d'Armiro qui mènent à Rettimo, et ont arboré leurs enseignes à deux milles de la Canée."

It would be easy to multiply these authorities, to shew how fearfully great was the provocation given by the Mohammedans to the unfortunate Christians of the island, both in the whole of the last century, and more especially during the few years which preceded the commencement of the Greek revolution. I will however content myself with citing here, in order to confirm the account of M. D'Herculez, by the testimony of another eye-witness, the words of the French Consul's official communication to his Government respecting the proceedings of the Khanióte Mohammedans in May, June, and July, 1821:

"*A la Canée, le* 24 *Juillet*, 1821. Le mois de Juin, époque du Ramazan, se passa jusq'au 28, sans autre meurtre que celui de l'infortuné Evêque de Kissamos, et du précepteur de l'enseignement mutuel. Le 29 la population turque fut armée avec les armes prises dans les arsenaux, et les enseignes de guerre furent arborées et promenées, dans tous les quartiers de la ville, par quatre bandes armées, qui donnèrent le signal du carnage medité depuis plus d'un mois. Un paisible forgeron Grec fut assassiné dans son atelier par la population armée: le 30 le drogman de la Porte fut aussi assassiné chez lui, un prêtre grec le fut aussi, et il suivit le pillage entier des vases sacrés, de l'argenterie, et de tous les ornemens sacerdotaux de l'église épiscopale grécque.

"Tous les Grecs trouvés dans leurs maisons, ou rencontrés dans les rues ou dans la campagne, furent immolés à la fureur des Turcs, sans distinction d'age et de sexe : leurs têtes furent coupées, et jettées dans les places publiques, et devant les casernes des régimens, où elles ont éte mutilées par la populace turque.

"*Les jours suivans les maisons grecques des villages circonvoisins, et plusieurs monastères, ont été incendiés et pillés; beaucoup de femmes et filles violées, et vendues ensuite, au bazar, comme esclaves. D'autres ont été assassinées, aussi que des enfans et des vieillards—Ceux qui ont été pris isolément ont été empalés ou pendus en ville—quoique ils eussent été pris sans armes. Enfin chaque jour du mois de Juillet a été marqué par des scènes d'horreur, et la campagne n'offre plus que le spectacle de la destruction et de la terreur.*"

Within the walls of Megálo-Kástron the carnage was far greater than at Khaniá. The venerable Metropolitan, five Bishops, four Priests, and more than seventy other individuals, had sought for refuge at the altar of the cathedral; but the brutal and infuriated populace rushed in upon them, and they were all sacrificed on the spot. The whole number of unarmed and innocent Christians massacred, within Megálo-Kástron, during the next few hours, was nearly six hundred: and, on the following day, an equal number were slaughtered in the districts of Malevízi, Pedhiádha, and Témenos[4].

Thus the Greeks had seen the most venerable and dignified ministers of their religion butchered at the very altar, or ignominiously dragged through the streets, and then hanged by the road-side: they had seen their fellow Christians, their relations and their friends, murdered while peaceable and unarmed: they had seen their wives and daughters violated, and then sold as slaves: and when, after all this, any unfortunate Mohammedan fell into the hands of the insurgents, we shall hardly wonder that his hours were numbered.

[4] As to what happened on this occasion at Venerâto, see above, Vol. I. p. 233.

ASKYFO.

CHAPTER XXXV.

WINE OF ASKYFO. SFAKIAN CHEESE-CAKES. CRETAN OATHS, ANCIENT AND MODERN. EUPHEMISM. DEPARTURE FROM ASKYFO. TREES OF THE SFAKIAN MOUNTAINS. ARRIVAL AT ANOPOLIS. DIALECTIC PECULIARITIES OF THE SFAKIANS. CRETANS AT ODESSA ON THE OUTBREAKING OF THE GREEK REVOLUTION.

May 5.

SEVERAL of the principal villagers beg of me to visit them, which I do soon after sunrise, accompanied by my guide Maniás. Roast lamb and excellent wine[1]

[1] This is the highest elevation at which the ground is cultivated in Crete, and the wine is most excellent. THEOGNIS, v. 875. speaks of wine produced on the summit of Taygetos:

Πῖν' οἶνον, τὸν ἐμοὶ κορυφῆς ἄπο Τηΰγέτοιο,
ἄμπελοι ἤνεγκαν.

are every where produced, and Maniás seems both willing and able to make up for the privations of Lent, by the quantity which he eats and drinks now. After this round of visits, my host would not hear of our departing without breakfasting with him, as if nothing had happenned elsewhere. Our fourth or fifth breakfast consisted of lamb, and fried cakes filled with a kind of cream-cheese, and which are a very general article of feast-day-food in Sfakiá. They are about three times as thick as a common English pancake, and of nearly the same superficial size. The Sfakians call them misethróftes[2], almost literally cheese-cakes. I do not remember to have seen them out of Crete. This kind of cake, whether peculiar to the Cretans or not, reminds me, that in ancient times there was a particular name for the thin cheeses made by the Cretans, and that they used to be offered in certain sacrifices[3].

On a subsequent occasion, when dining at Askýfo, in the house of the celebrated Christian leader Rússos, I was much struck with the variety and singularity of the oaths which he used: I greatly regret that I did not note them all down at the time. One was "by the bread which I am eating[4]," another "by my father's bones[5]." In these singularly varied oaths the name of the Deity was always avoided: and, even in the very common Cretan expression *μὰ τὸν ἐς*, which has been already explained[6], a similar feeling may be traced.

It appears quite clear that the same peculiarity was observed in the oaths of the ancient Cretans: the difference between their oaths and those of the other Greeks, was remarkable enough to give rise to a legend, that Rhadamanthus had especially charged the Cretans,

[2] Μισηθρόφταις.

[3] ATHENAEUS, XIV. p. 658. d. (= p. 1465. ed. Dind.) Τοὺς δὲ λεπτοὺς τῶν τυρῶν καὶ πλατεῖς Κρῆτες θηλείας καλοῦσιν, ὥς φησι Σέλευκος· οὓς ἐν θυσίαις τισὶν ἐναγίζουσιν.

[4] Μὰ τὸ ψωμὶ ἀποῦ τρώγω.

[5] Μὰ τὰ κόκκαλα τοῦ κύρού μου.

[6] Above, Vol. I. p. 249.

"that they might avoid naming the Deity on every occasion, to swear by the goose, and the dog, and the ram[7]." Every one will remember the custom of Socrates.

Another euphemism attracted my attention, at Askýfo, in an expression of my hostess. She used a word, which I had never before heard, to denote the Devil, while she avoided pronouncing his ordinary name[8].

We do not leave Askýfo till half-past eight: after crossing the plain we begin an ascent, which continues, with but little intermission, till twenty minutes before ten, when I find snow, of which my horse eats greedily, lying three feet thick on the side of the road. After a slight descent, which is succeeded by an ascent, I begin, at half-past ten, to follow the windings of a very bad road along the south side of this great chain of the White Mountains, and not very far from their snow-clad summits. Soon after eleven the road improves, and changes its direction to the south-west, the African sea and the island of Gávdha being in sight.

At half-past eleven we arrive at a fountain of deliciously cold water, at which is a solitary fig-tree. Trees grow on all these rocky mountains, except quite on the summits of the highest ranges. The commonest tree is the prínos, or ilex, and, near the fountain, are also a great many cypresses. We are less than two miles to the north-north-west of the village of Murí. At half-past twelve we emerge from a valley, which we have been following for some time, and see Gávdha and

[7] EUSTATHIUS, on Odyss. T. p. 1871. 3. Ῥαδάμανθυς δέ, φασιν, ὑπὲρ τοῦ μὴ Θεὸν ὀνομάζειν ἐπὶ πᾶσιν ἐκέλευσέ, φασι, κατὰ χηνός, καὶ κυνός, καὶ κριοῦ ὀμνύναι. See also SUIDAS under Ῥαδαμάνθυος ὅρκος, and Porphyry, Zenobius, and Apostolius, all quoted by MEURSIUS, Creta, p. 195.

[8] Ὁ τόπιος, which I suppose must mean "He who lives in *the* place," κατ' ἐξοχήν. She said, "You should say so, θὰ μὴ βλασφημήξῃς—θὰ 'πῇς καρὰ τὸ ὄνομά του—τοῦ διαβόλου." To give another instance or two of the Sfakian change of λ into ρ, I may quote a few more words from the pretty mouth of my hostess. "When a man gets married, σκραβώνεται," (for σκλαβώνεται). In speaking to her child, she says, ἔρα, ἔρα! instead of ἔλα, ἔλα! The καθέκλα is changed into καθέκρα.

Ghávdha-púla, as well as the Paximádhia islands, and the projecting point of Mesará: as I look back, both Psylorítes and Kendros are now in sight. Murí is about a mile to our south-east and Anópolis is four miles to the south-west. From this point we continually descend along the side of the mountains, by a very bad road, towards the plain of Anópolis, and we reach its northern edge at half-past one.

The villages designated by the common name Anópolis, are Limnaéa, Skála, Mariană, Gýros, Kámpos, Hághio Dhemétrios, Kampiá, and Rhíza, at the last of which we arrive after crossing the plain. Rhíza is situated on the slope of a rocky elevation on the south side of the plain, which, small as it is, is not entirely cultivated. On entering this hamlet we found a cluster of Greeks, many of whom were, as might have been expected, friends of Captain Maniás. They were all idling away the day in honour of its being Easter Monday, and the salutation "Christ is risen," with its appropriate reply, and the subsequent embrace, has to take place a score times, between Maniás and his friends, before we reach our quarters, the house of an Anopolítan whom I had met near Askýfo, to which place he was going, and who warmly pressed me to lodge with him, and to accept of such hospitality as he could shew me. My host returned to his home before nightfall. As I hear his wife talk, I cannot help again noticing the marked peculiarities of dialect which characterize all the Sfakians, especially the women, and some of which may undoubtedly be considered as traces of the old Cretan-doric.

My discovery of these peculiarities in the language of the Sfakians has given me great pleasure. No one, either traveller or resident in the cities, seems to have been aware of their existence. The Sfakians were described to me at Khaniá as ferocious savages, who spoke a still more corrupt dialect than the other Cretans; while the fact is, that they speak a language which is almost purity, when compared with the jargon of a

Smyrniote[9], or of the inhabitant of any considerable maritime town, except Constantinople, throughout the whole of the Turkish empire.

The frequent substitution of ρ for λ, and the formation with ξ of a greater number of aorists than is found elsewhere, are the most striking of these peculiarities[10], and are likewise the most interesting, since they are obviously derived from the old language of the island: and this is perhaps the only district inhabited by Greeks where such clear traces of an ancient *dialect* are discernible[11]. I am, however, told by Prince Mavrokordáto, that he has observed vestiges of the ancient Doric in the language of Icaria, an island which has had but little intercourse with any other place.

These remarkable archaisms are most conspicuous, as I have said, in the language of the Sfakian women, and thus call to mind the words of Cicero: "Facilius mulieres incorruptam antiquitatem conservant, quod multorum sermonis expertes, ea semper tenent quae prima didicerunt[12]."

My hostess tells me that the ancient remains, of which I had first heard when in the eparkhía of Sélino, are at the top of the rocks, near the foot of which this village is situated, and from which it derives its name. She says that the sea[13] is visible from them.

In ancient times, as I learn from the villagers, the city below at Lutró[14], was in alliance with this above,

[9] To mention a few words out of scores which might be adduced, the Sfakian peasant would say, *στρατεύματα* or *δυνάμεις*, *διαφορά*, *παραθύρι*, while the Smyrniote merchant would use the Italian words *τρούππαις*, or *σολδάταις*, *διφφερέντζα*, *φινέστρα*, &c.

[10] Others will be observed, such as the frequent addition of a final ι at the end of past tenses, and words which are used by the Sfakians alone.

[11] As to the Dorisms of the modern Tzakonic dialect, the reader may consult Colonel LEAKE's Travels in the Moréa, Vol. II. pp. 505—508. and his Researches in Greece, p. 66. and p. 197. foll.

[12] CICERO, De Orat. III. 12. where Ursinus quotes from the Cratylus: *Τυναῖκες μάλιστα τὴν ἀρχαίαν φωνὴν σώζουσι.* See also the book de Claris Oratoribus, c. 58. and HARRIS's Philosophical Inquiries, p. 458.

[13] Ἡ *θάρασσα.*

[14] Εἰς τὸ Ῥουτρό.

and was called Katópolis, as this is called Anópolis. "The two possessed between them no less than seventy-two thousand troops[15]." This modern village is a most healthy place: "no one is ever known to be ill here." My hostess's father died last year at eighty-five. "He knew all about Lambro's[16] expedition, and understood more than other men." No one would have believed him to have been so old. Four years before his death he became blind[17]. During the revolution she fled with her two children, her husband remaining always in Crete, first to Gávdha, then to Pholékandhros, and then to Mélos[18].

I am assured, that an ancient stone, such as is sometimes found in a ring, whether engraved or not, performs the office of a propitious Ilithyia to women in labour[19].

One of the inhabitants of the village, who accompanied me to examine the ancient remains which are found near it, was at Odessa, at the outbreaking of the revolution, with about a hundred and forty other Greeks, ten of whom were Cretans and three Sfakians. They were all sailors, on board different vessels then in the port of Odessa, and knowing that their lives would be forfeited if they attempted to pass Constantinople on their return to Greece, they went by land to Marseilles, where they embarked for Hýdhra. The policy of Prince Metternich refused to allow a soul of them to enter the Austrian dominions, and they therefore had to traverse Russia, Poland, Northern Germany, and Switzerland, to Lyons and Marseilles. When at Warsaw they had an interview with the Grand Duke, who spoke Greek quite fluently, and, having heard that there were some Sfakians among them, asked which they were, and addressed his

[15] 'Εβδομῆντα δύω χιλιάδες ἀσχέρι.

[16] Lambro was called 'Ράμπρο by all of them.

[17] 'Ετυφλώθη καὶ δὲν ἔβλεπε were her words.

[18] Εἰς τὴν Μήρον.

[19] If after childbirth the woman is able to nurse her own infant, the stone has another valuable property, καταβάζει τὸ γάρα. (τὸ γάλα.)

conversation particularly to my informant and his fellow Sfakians. The Russian government paid all their travelling expenses, and even supplied them with money. So did most of the governments in the other countries through which they passed.

CRETAN FEMALE PEASANT.

CHAPTER XXXVI.

COSTUME OF THE CRETAN WOMEN SOMEWHAT CHANGED SINCE TOURNEFORT'S TIME. THE KATAKHANAS, VURVULAKAS, OR VAMPIRE. A WELL-ATTESTED CRETAN VAMPIRE-STORY. THE VAMPIRE FEEDS ON THE HUMAN LIVER. THE PRIESTHOOD SUPPOSED TO HAVE THE POWER OF EXORCISING A VAMPIRE. THE POPULAR NOTIONS NOW ENTERTAINED, RESPECTING VAMPIRES, IN GREECE, DALMATIA, AND ELSEWHERE, WERE FORMERLY PREVALENT IN ENGLAND. ETYMOLOGIES OF THE WORDS VURVULAKAS AND

KATAKHANAS. A BELIEF IN VAMPIRES ENTERTAINED BY THE JEWS AND ARABS; AND ALSO FOUND IN SANSKRIT LITERATURE. SIMILAR NOTIONS OF THE ANCIENT GREEKS. THE NEREIDHES, OR "BEAUTIFUL LADIES," OF MODERN GREECE, RESEMBLE BOTH THE NEREIDS OF ANCIENT GREECE, AND THE FAIRIES OF OLD ENGLAND. OTHER LEGENDS. A MID-DAY DEMON OF THE MOUNTAINS. THE POWER OF WITCHCRAFT SUPPOSED TO BE POSSESSED BY THE CLERGY. DIFFERENT POSITION OF THE CLERGY IN GREECE AND ENGLAND WITH RESPECT TO WITCHCRAFT. ADDITIONAL NOTES ON VAMPIRISM, AND ON THE ORIGIN OF THE ENGLISH CUSTOM OF DRIVING A STAKE THROUGH THE BODIES OF SUICIDES. SPECIMENS OF THE SFAKIAN DIALECT.

THE costume of the women in Crete has undergone some little alteration since the time of Tournefort. The trowsers which they now wear in every part of the island, were nowhere seen by him. "Les dames de l'Archipel portent des caleçons: les Candiotes n'ont que la chemise sous leur jupe."

The French naturalist seems to have admired the rarities of the inanimate more than those of the animate creation, in Crete. He says, "A l'égard des dames nous en avons veû d'assez jolies à Girapetra: ailleurs elles sont laides." He thus describes their dress: "Cet habit est très simple: C'est une jupe de drap rouge tirant sur le grisdelin, fort plissée, suspendue sur les épaules par deux gros cordons, et qui leur laisse le sein tout découvert."

When at Askýfo I had asked about the Vampires, or Katakhanádhes as the Cretans call them, of whom I had heard from Maniás and others of his fellow-countrymen, and whose existence and ill-deeds form a general article of popular belief throughout the island. Of course this belief is very strong in the mountains. If any one ventures to doubt it, undeniable facts are brought forward to silence the incredulous. At Anópolis I am on ground which has long been haunted by them, and is celebrated in numerous stories, some of which are amusing enough, in which their exploits are recorded.

I subjoin[1] one of these stories in the very words in which it was communicated to me. The account is peculiarly worthy of credit, since I heard it in many places, and all the relations given to me agreed in every material point. The following is a translation, and, even without comparing it with the original, the reader will see, from its very style, that it is a close, though somewhat condensed, version of the words of the Sfakian peasants.

"Once on a time the village of Kalikráti, in the district of Sfakiá, was haunted by a Katakhanás, and people did not know what man he was or from what part. This Katakhanás destroyed both children and many full-grown men; and desolated both that village and many others. They had buried him at the church of Saint George at Kalikráti, and in those times he was a man of note, and they had built an arch over his grave. Now a certain shepherd, his mutual Sýnteknos[2],

[1] See the original Sfakian Greek in Note A, at the end of the Chapter.

[2] I believe I first heard this word in Crete: it is always used to denote the relation of a person to his god-child's father. The one is the spiritual, the other the natural father of the same child: hence they are well called *Σύντεκνοι*. Here as in many other instances the simple word (*τέκνον*) has gone out of use, although the compound is still retained. See Vol. I. p. 63. Sýnteknos is more common, among the Cretans, than *compare* (*κουμπάρης*), the word generally used elsewhere to express this relation. On *compater* and *commater* see Du Cange, Glossarium med. et inf. Lat. under *Commater*, *Commaternitas*, *Compater*, *Compaternitas*, and *Compateratus*. The mother of a man's god-child is also his *σύντεκνος* or *κουμπάρισσα* (the diminutive of which latter word is *κουμπαρισσάκι*). Our Cretan word ΣΥΝΤΕΚΝΟΣ occurs, in its present sense, in a sepulchral inscription which was dug up near the Savoy in London, three or four years ago, and seems to have belonged to the collection formed by the Earl of Arundel, in the early part of the seventeenth century. See the Transactions of the Royal Society of Literature, Vol. II. Part II. Appendix I. pp. 463–4. The expression, mutual Sýnteknos, is used, as my Sfakian companion explained it, for persons *ἀποῦ εἶχε παιδὶ ὁ γῆς τοῦ ἄῤῥου βαπτισμένου*, where *ἀποῦ* stands for *ὁ ὁποῖος*, *ὁ γῆς* for *ὁ ἕνας*, and *ἄῤῥου* for *ἄλλου*. On this gossipred, see Vol. I. pp. 10–11. where I promised to recur to the subject. The relation is considered, in the Greek Church, as complete a bar to marriage as the closest consanguinity. A man could never wed a widow, if he had been sponsor for any of her children at the baptismal font; and a Greek would almost as soon think of marrying the daughter of his own father as the daughter of his god-father.

was tending his sheep and goats near the church, and, on being caught by a shower, he went to the sepulchre, that he might be shaded from the rain. Afterwards he determined to sleep, and to pass the night there, and, after taking off his arms[3], he placed them by the stone which served him as his pillow, crosswise. And people might say, that it is on this account[4] that the Katakhanás was not permitted to leave his tomb. During the night, then, as he wished to go out again, that he might destroy men, he said to the shepherd: "Gossip, get up hence, for I have some business that requires me to come out." The shepherd answered him not, either the first time, or the second, or the third; for thus[5] he knew that the man had become a Katakhanás, and that it was he who had done all those evil deeds. On this account he said to him, on the fourth time of his speaking, "I shall not get up hence, gossip, for I fear that you are no better than you should be, and may do me some mischief: but, if I must get up, swear to me by your winding-sheet, that you will not hurt me, and on this I will get up." And he did not pronounce the proposed words[6], but said other things: nevertheless, when the shepherd did not suffer him to get up, he swore to him as he wished. On this he got up, and, taking his arms, removed them away from the monument, and the Katakhanás came

father. HARTLEY, Researches in Greece, p. 79. "A young man of Ithaca informed me, that it was difficult to obtain a wife in his native island; for the principal families had become connected with each other, to such an extent, by *marriages* and *baptisms*, that it was almost necessary to resort to Cephalonia or elsewhere, in order to effect a marriage."

[3] Arms were part of every Sfakian's dress, and were never laid aside by him except for the hours during which he slept.

[4] The readers must not suppose that the Vampire feared the mortal weapons: no, they were placed *crosswise*, and, therefore, the form of the cross, and not the cold steel, repelled him. Some people say that the cross made by the handle of the sword produced this effect: others believe it was owing to one of the pistols having been laid across the other.

[5] That is, by the Vampire's words, which betrayed his inability to come forth so long as his pious friend and his crossed arms were lying upon him.

[6] That is, he did not pronounce the words of the only oath which binds a Katakhanás: μὰ τὸ ἀναβόλι μοῦ. "By my winding-sheet."

forth, and, after greeting the shepherd, said to him, "Gossip, you must not go away, but sit down here; for I have some business which I must go after; but I shall return within the hour, for I have something to say to you." So the shepherd waited for him.

And the Katakhanás went a distance of about ten miles, where there was a couple recently married, and he destroyed them. On his return, his gossip saw that he was carrying some liver, his hands being moistened with blood: and, as he carried it, he blew into it, just as the butcher does, to increase the size of the liver. And he shewed his gossip that it was cooked, as if it had been done on the fire[7]. After this he said, "Let us sit down, gossip, that we may eat." And the shepherd pretended to eat it, but only swallowed dry bread, and kept dropping the liver into his bosom. Therefore, when the hour for their separation arrived, the Katakhanás said to the shepherd, "Gossip, this which you have seen, you must not mention, for, if you do, my twenty nails will be fixed in your children and yourself[8]." Yet the shepherd lost no time, but gave information to priests, and others, and they went to the tomb, and there they found the Katakhanás, just as he had been buried. And all people became satisfied that it was he who had done all the evil deeds. On this account they collected a great deal of wood, and they cast him on it, and burnt him. His gossip was not present, but, when the Katakhanás was already half consumed, he too came forward in order that he might enjoy the ceremony. And the Katakhanás cast, as it were, a single spet of blood, and it fell on his foot, which wasted away, as if it had been roasted on a fire. On this account they sifted even the

[7] GUILIELMUS ALVERNUS, p. 1066. quoted by J. GRIMM, Deutsche Mythologie, p. 596. "Idem, et eodem modo sentiendum est tibi de aliis malignis spiritibus, quos vulgus *stryges* et *lamias* vocat, et *apparent de nocte in domibus* in quibus parvuli nutriuntur, eosque de cunabulis raptos, laniare vel *igne assare* videntur."

[8] Meaning, I suppose, that he should destroy them, and carry off their liver, in the same manner as he had done that of the newly-married couple.

ashes, and found the little finger-nail of the Katakhanás unburnt, and burnt it too[9]".

This supposed Vampire's habit of feeding on the human liver, may perhaps account for an exclamation of a Cretan mother, recorded in the travels of Tavernier: "I will sooner eat the liver of my child[10]."

The Vampire, or Katakhanás, as he is called in Crete, is denominated Vurvúlakas, or Vrukólakas, in the islands of the Archipelago, where the belief is generally prevalent, that if a man has committed a great crime, or dies excommunicated[11] by a priest or bishop, the earth will not receive him when he dies, and he therefore rambles about all night, spending only the day time in his tomb. Many believe that, even in the day time,

[9] I hear other Katakhanádhes spoken of at Anópolis as having made terrible ravages: *Καὶ προτήτερος ἄῤῥος ἕνας—καὶ ὅσους κἂν ἀποθάνασι τζὴ ἐπέρνα. καὶ ἐγυρίσασι (ἄῤῥοι Καταχανάδες) καὶ ἐπηράσασι πορῤοὺς ἀνθρώπους.*

[10] TAVERNIER, Six Voyages, Liv. III. Ch. VIII. p. 310. ed. Paris, 1676. "La femme, n'y voulant consentir, luy dit *qu'elle mangeroit plû-tost le foye de son enfant*, que de satisfaire à son infame désir." I find a similar expression in the poem of KHADGI SERET: LEAKE, Travels in Northern Greece, Vol. I. p. 492.

Ἀλῆ-πασιᾶς ἐγύρισε
καὶ ἐχαΐδευσε τὸ Φώτη,
καὶ τήραξε, τοῦ εἶπε, καλά,
τὶ ἐγώ σου καίω τὸ σκότι.

Alì-pashâ Fôto caressed
As then he round did turn,
And "Look well to it," he exclaimed,
"*Lest I thy liver burn.*"

[11] *Γίνεται ἀφορισμένος ἀπὸ τὴν πίστιν του.* Several forms of excommunication, in which the offender is condemned to remain *μετὰ θάνατον ἄλυτος αἰωνίως ὡς αἱ πέτραι καὶ τὰ σίδηρα*, are given by CHRISTOPHORUS ANGELUS, RICAUT, and GOAR, (on the Greek Euchology, p. 688.) all cited by J. M. HEINECCIUS, de absolutione mortuorum excommunicatorum seu tympanicorum in Ecclesia Graeca, (Helmstad. 1709.) pp. 10—13. The ministers of religion, who claimed the power of thus preventing a body from returning to dust, naturally held, that when *they* absolve the dead offender, it ought at once to become decomposed. It is said that this used always to be the case. A remarkable instance of the exercise of this power is quoted, (of which Mohammed the Second is said to have been a witness,) by CRUSIUS, Turco-Graecia, II. pp. 27-28. GOAR, Euchelog. Graec. p. 689. ALLATIUS, l. c. BZOVIUS, in the year 1481, n. 19. and other authors cited in J. M. HEINECCIUS, De absolutione, &c. p. 16.

it is only once a week, on the Saturday, that he is allowed to occupy his burial-place. When it is discovered that such a Vurvúlakas is about, the people go, on a Saturday, and open his tomb, where they always find his body just as it was buried, and entirely undecomposed. The priest by whom they are accompanied reads certain parts of the ritual, supposed to be of peculiar efficacy for putting a stop to every restless Vampire's wanderings, and sometimes this course suffices to restore the neighbourhood to peace and quiet. But cases happen in which the priest is not a sufficiently powerful exorcist, thus easily to stop the nocturnal rambles and misdeeds of the undying one, who, like Shakspeare's ghost, is doomed to walk the night, as a punishment for the foul crimes done in his days of nature. Whenever, then, this ordinary religious ceremony, to which recourse is first had, is found inefficacious, the people of the neighbourhood go to the tomb on a Saturday, take out the body, and consume it with fire; an operation which nothing but extreme necessity would ever make Greeks consent to perform, on account of their religious horror of burning a body on which the holy oil has been poured by the priest when performing the last rite of his religion over the dying man.

Even the rough Hydhraeans[12], whose seafaring life and intercourse with other countries, might have been supposed likely to have diminished the prevalence of such notions among them, are generally believers in these Vurvúlaki. As in Sfakiá, so also at Hýdhra,

> Both well attested, and as well believ'd,
> Heard solemn, goes the *Vampire*-story round;
> Till superstitious horrour creeps o'er all.

Many Hydhraeans have assured me there used to be a great number of Vampires in Hýdhra, and that

[12] Οἱ Ὑδραῖοι, who are commonly called Hydhriotes by European writers: I have adopted the name by which they are known to all who speak Greek.

their present freedom from them is to be attributed solely to the exertions of their bishop, who has laid them all in Santoréne, where, on the desert isle, they now exist in great numbers, and wander about, rolling stones down the slopes towards the sea, "as may be heard by any one who passes near, in a kaík, during the night".

The Sfakians also generally believe that the ravages committed by these night-wanderers, used, in former times, to be far more frequent than they are at the present day; and that they are become comparatively rare, solely in consequence of the increased zeal and skill possessed by the members of the sacerdotal order!

Similar exertions of holy priests are said, by our poet Chaucer, to have put a stop to the pastimes of the English fairies[13]. Thus also Dryden:

Lo in the walks where wicked elves have been,
The learning of the parish now is seen—
From fiends and imps he sets the village free,
There haunts not any incubus but he:
The maids and women need no danger fear
To walk by night, and sanctity so near.

[13] After telling us that,

In olden dayes of the king Artour,
All was this land fulfilled of faerie,

he adds,

But now can no man see non elves mo,
For now the grete charitee and prayeres
Of limitoures, and other holy freres
That serchen every land, and every streme,
As thicke as motes in the sonne beme;
This maketh that there ben no faeries.

A later bard, the author of "The Fairies' Farewell," attributes their disappearance to the Reformation:

By which we note the fairies
Were of the old profession;
Their songs were Ave Maries,
Their dances were procession.

But now, alas! they all are dead,
Or gone beyond the seas;
Or farther for religion fled,
Or else they take their ease.

See SCOTT's edition of Dryden's Works, Vol. XI. p. 378.

This popular belief in Vampires is not confined to Crete and Greece[14]; but, as most of my readers will be aware, is very widely spread: they are found in Dalmatia[15], Hungary, Moravia, and other countries[16].

During the middle ages, a belief in ravages committed by similar monsters, was not confined to poets, in our own island, but formed an article of generally prevalent superstition among the people, and, like the equally absurd belief in witchcraft of more recent times, was shared with them by the clergy.

William of Newbury, who flourished in the twelfth century, relates that, in Buckinghamshire, a man appeared several times to his wife, after he had been buried. The aid of the church was called in, as is still done on similar occasions by the Greeks. The archdeacon and clergy thought it right to apply to the bishop (of Lincoln), who learnt that such events were of frequent occurrence in England, and that the only known remedy was to burn the body[17], which, on opening the grave, was found in the same state as on the day of its inter-

[14] Tournefort has given a long account of the proceedings of one of these blood-sucking nocturnal monsters, at Mýkonos, during the time of his stay in that island: TOURNEFORT, Voyage du Levant, Tom. I. Lettre III. pp. 131—136. He calls it a *Vroucolakas*.

[15] FORTIS, Travels into Dalmatia, l. c.

[16] See DOM CALMET, Dissertations sur les Apparitions des Anges, des Démons, et des Esprits, et sur les Revenans et Vampires de Hongrie, de Bohème, de Moravie et de Silesie, Paris, 1746, pp. 261—303. HORST'S Zauberbibliothek I have not been able to see.

[17] GUILIELMUS NEUBRIGENSIS, Chronica rerum Anglicarum, Lib. V. c. XXII. p. 569. ed. Hearn. Oxon. 1719. "Talia saepius in Anglia contigisse, et crebris clarere exemplis, quietem populo dari non posse, nisi miserrimi hominis corpore *effosso et concremato*." As to the burning, even Turks have had recourse to this remedy to put at rest a Christian vampire. CRUSIUS, Turco-Graecia, VII. p. 490. "In sabbato pentecostes Turcae combusserunt Graecum, biennio ante defunctum: quod vulgo crederetur noctu sepulcro egredi, hominesque occidere. Alii autem veram causam perhibent, quod quindecim pluresve homines spectrum ejus videntes mortui sint. Sepulcro extractus, consumpta carne cutem ossibus adhaerentem integram habuit." Compare HEINECCIUS, l. c. p. 20.

ment[18]. The same author mentions a similar story, the locality of which was the town of Berwick-upon-Tweed, where the body was cut in pieces and burnt[19]. Another Vampire was burnt at Melrose Abbey. He had been a very worldly priest[20], and so fond of hunting that he was commonly called *Hundeprest*. A still worse case occurred at a castle in the north of England, where the Vampire so frightened all the people, that no one ever ventured out of doors between sunset and sunrise, "Ne forte oberranti monstro sugillandus occurreret." The sons of one of his supposed victims at length opened his grave, and pierced his body, from which a great quantity of blood immediately flowed, and plainly proved how many persons had been his victims[21].

[18] "Aperto ergo sepulcro, corpus ita inventum est, sicut ibidem fuerat collocatum." Our ancestors in the reign of Henry III. learnt how a bishop's excommunication could thus preserve a dead body from dissolution; but the inflexible sternness of the prelate's character, and the heavy price at which he rated his forgiveness, prevented the exhibition of the effects of episcopal absolution: see MATTHEW PARIS, Hist. Anglic. p. 666. and J. M. HEINECCIUS, De absolutione, &c. pp. 26–27.

[19] "Conduxerunt decem juvenes audacia insignes, qui corpus infandum effoderent, et membratim exectum redigerent in combustionem."

[20] "Supra modum secularis extiterat."

[21] "Vulnus exanimi corpori intulerunt: ex quo tantus continuo sanguis effluxit, ut intelligeretur sanguisuga fuisse multorum." These blood-suckers or incubuses were commonly supposed, in England, to be the Devil or some of his chief ministers. MILTON calls Belial (P. R. II. 150.)

> The sensuallest, and after Asmodaï
> The fleshliest Incubus.

SHAKSPEARE, Tempest, Act I. Sc. II. gives Caliban the same Arch-Incubus as his sire. We find that in order to seduce English *men* the demons used to assume the form of *women*, and were called *succubuses*, or *succubae*. In THOMAS MIDDLETON, A mad world my masters, Act IV. (in Dodsley's Old Plays, Vol. V. p. 329.) the virtuous country gentleman is addressed by a Succubus:

> Have I this season wrought with cunning,
> Which, when I come, I find thee shunning?
> When was place and season sweeter?
> Thy bliss in sight, and dar'st not meet her?

And in Todd's Johnson's Dictionary, there are quotations, in which are mentioned, an ancient grandame, who

> Though seeming in shape a woman natural,
> Was a fiend of the kind that succubae some call;

and

With these narrations, especially the last, the Hungarian legends collected by Dom Calmet[22], should be compared.

The enlightenment of the episcopal order in some parts of Turkey, at the present day, would seem, from the following story, to be greater than that of our own English Bishops seven hundred years ago. About the commencement of the present century, the Metropolitan of Larissa received advice of a papás having disinterred two bodies, and thrown them into the Haliacmon, on pretence of their being Vrukólakas. "Upon being summoned before the bishop, the priest confessed the fact, and asserted in justification, that a report prevailed of a large animal having been seen to issue, accompanied with flames, out of the grave in which the two bodies had been buried. The bishop began by obliging the priest to pay him two hundred and fifty piastres; (his holiness did not add that he made over the money to the poor.)—By then publishing throughout the diocese, that any similar offence would be punished with double the fine and certain loss of station, the bishop effectually quieted all the vampires of his episcopal province[23]."

and "a church-yard carcass, raised and set a strutting by the inflation of some hellish succubus within." The entrance of a demon into the dead body is also mentioned, as a part of the modern Greek superstition, by Leo Allatius, Ricaut, and other authors.

[22] Calmet, Dissertation etc. p. 275. foll. One is mentioned at p. xx. of the Introduction to the Vampyre, a tale, London, 1819, from the London Journal for March, 1732. See also the story told in Ricaut, (State of the Greek Church, p. 278. foll.) which is highly illustrative of the superstition. The Morlacchians seem to have found the burning or boiling described by Calmet and Ricaut unnecessary: "When a man dies suspected of becoming a Vampire or Vukodlak, as they call it, they cut his hams, and *prick his whole body with pins*, pretending that, after this operation, he cannot walk about. There are even instances of Morlacchi, who, imagining that they may possibly thirst for childrens' blood after death, intreat their heirs, and sometimes oblige them to promise, to treat them as vampires when they die." Fortis, l. c. See some further researches, connected with this subject, in the Note B, at the end of the Chapter.

[23] Leake, Travels in Northern Greece, Vol. iv. p. 216.

In the greater part of civilized Europe, these vampire-stories are now found only in the language of poetry. A "vampire corse" is introduced by Southey, in "Thalaba the Destroyer[24]," and Byron has alluded to the superstition in his Giaour:

> But first on earth as vampire sent
> Thy corse shall from its tomb be rent:
> Then ghastly haunt thy native place,
> And suck the blood of all thy race.

In Goethe's Bride of Corinth[25], a young Athenian visits a friend of his father's, to whose daughter he had been betrothed, and is disturbed at midnight by the appearance of the vampire spectre of her whom death had prevented from becoming his bride. She addresses him:

> Aus dem Grabe werd' ich ausgetrieben,
> Noch zu suchen das vermisste Gut,
> Noch den schon verlohrnen Mann zu lieben,
> Und zu saugen seines Herzens Blut.

A legend, related in the second volume of the Minstrelsy of the Scottish Border, has also served as the groundwork of a poem, "The Vampire Bride," in which at last the vampire's grave is opened, and

> When they thrust a dart through its swollen heart
> It convulsively shivered and screamed[26].

Traditions of this kind have also been made the foundation of a horrible story, called "The Vampyre, a Tale," published at London, in 1819.

The vampire in all these cases, and in most of the popular legends, is described as a blood-sucking monster, not unlike the insatiable wolf of Dante, which

> Ha natura si malvagia e ria,
> Che mai non empie la bramosa voglia.

[24] See SOUTHEY'S Thalaba, VIII. note at p. 108. ed. of 1814.

[25] GOETHE, Die Braut von Corinth, among his "Balladen," Werke, I. Bd. pp. 221—228. ed. Stuttg. und Tuebing. 1827.

[26] LIDDELL'S Vampire Bride, p. 52. ed. Edin. and Lond. 1833.

We must now return to the Vurvúlakas of the Egean, and to the Katakhanás of Crete.

Koray supposes the word Vurvúlakas to be derived from the ancient Greek[27]. I find, however, that the Morlacchians[28], the inhabitants of Montenegro[29], the Bohemians, and the Arnauts, both of Hýdhra and Albania[30], call these vampires by the same name Vurvúlakas, or by one evidently springing from a common source with it; and I should therefore suppose the word to be of Slavonian rather than of Hellenic origin. This supposition is confirmed by our finding, that in Crete, where Slavonian influence has not been felt, as it has in the Moréa and in some islands of the Egean[31], the vampire is known by the totally different name of Katakhanás, a word which, though it may originally have meant simply

[27] KORAY, on Heliodorus, Tom. II. p. 5. and p. 199. and ATAKTA, Vol. II. p. 84. On the ancient μορμολύκη (from which Koray derives βουρβούλακας), and similar words, see HEMSTERHUIS, on Pollux, x. 167. p. 1353. RUHNKEN, on the Lexicon of Timaeus, p. 130. 31. ALBERTI, on Hesychius, Vol. II. p. 620. 21. Attempts which have been made to deduce βουρκόλακας from a Hellenic source, will be found in Note B, at the end of this Chapter.

[28] ALBERTO FORTIS, Travels into Dalmatia, Eng. tr. Lond. 1778. p. 61. He writes the word Vukodlak.

[29] VIALLA DE SOMMIERES, Voyage au Montenegro, Vol. I. ch. XXI. p. 279. Rien n'égale la terreur que leur inspirent les *brucolaques*, c'est-à-dire les cadavres des individus frappés d'excommunication, etc.

[30] I conversed more than once with Hydhraeans on this subject: as to the Epirotes, I find the word written βουρκόλακας in KHADJI SERET's Alí-pashá, (already quoted at p. 200.) a MS. poem in the possession of Col. Leake, who has given an account of it, interspersed with many extracts: see LEAKE, Travels in Northern Greece, Vol. I. pp. 463—497. At p. 492. we read,

εἶχε ἁμαρτιάς ὁ πικρός
  ἀπὸ τὰ γονικά του,
καὶ καίγουνταν βουρκόλακας
  στὰ γεροντάματά του.

For, on his head, full many a sin
  Descended from his sire,
And therefore he, when old, was burnt
  Like Vurkolak in fire.

The allusion is, I think, manifestly to the monk Samuel's death at Kughni, where he was blown up: see FAURIEL, Chants populaires de la Grèce moderne, Tom. I. p. 274. and LEAKE, Travels in Northern Greece, Vol. I. p. 240.

[31] See above, p. 82.

a destroyer[32], yet seems to have its peculiar Cretan signification even in the very poem where it occurs in the more general and earlier sense[33].

A belief in the existence of similar night-wandering blood-sucking monsters prevailed even in ancient times, and seems to have been one of the superstitions of the Jews. The *lilith*, mentioned in Isaiah, is agreed to be a nocturnal spectre: Michaelis supposes that it may mean a kind of incubus: and the Arabic version translates the word by "*algol*," the man-devouring demon of the waste, known by the name Goule to the English reader of the Arabian Nights. Some persons have supposed, but, seemingly, without any good reason, that the expression in the Psalms, "Thou shalt not be afraid for the terror by night, nor for the pestilence that walketh in darkness[34]," denotes a similar monster[35].

I have also been taught, by a communication recently received from a learned friend, Professor Von Bohlen

[32] EMMANUEL GEORGILLAS, in his poem entitled τὸ θανατικὸν τῆς Ῥόδου, quoted by KORAY, ATAKTA, Vol. II. p. 186.

νὰ πολεμίζουν δυνατὰ καὶ νὰπαντοῦν τὸν σκύλον
τὸν Τοῦρκον τὸν Καταχανάν, τῆς ἀνομίας τὸν φίλον.

The same poet, in his lament of Constantinople, (Θρῆνος τῆς Κωνσταντινουπόλεως,) in KORAY, Vol. II. p. 113.

Τὸν Τοῦρκον τὸν Καταχανὰν αὐτὸν τὸν χονικιάρην.

On which passage Koray explains the word as follows: Σημαίνει τὸν ἐξολοθρευτὴν ἀπὸ τὸ Καταχάνω, ἤγουν Καταχάόνω, ἐπειδὴ τὸ ἁπλοῦν καὶ σύνηθες Χάνω εἶναι, ὡς ἐσημειώθη—see ATAKTA, Vol. I. p. 267.

[33] GEORGILLAS in his Θρῆνος, as quoted by DUCANGE, Append. ad Glossarium. med. et inf. Graecit. p. 185. v. Τζακαρόλος, and by KORAY, ATAKTA, Vol. II. p. 185.

καὶ δένειν ἡ κακὴ ἀρχὴ αὐτὸς ὁ τζακαρόλος
και γίνεται Καταχανὰς καὶ ἀπέκει κάγην ὅλος.

In which passage KORAY writes δένεν and γίνετο. He conjectures that τζακαρόλος may be the Italian *tangherello*, and rightly explains κάγην as ἐκάη. He seems to have had no idea whatever of the meaning which the Cretan usage of the word shews to be contained in the passage.

[34] PSALM, XCI. 5, 6.

[35] BOCHART, Hierozoicon, P. II. L. VI. c. IX. p. 829. ed. Lips. 1796. Compare VAN DALE de Orig. et Prog. Idol. Diss. I. cap. VI. p. 131. On the absurd Rabbinical traditions concerning demons, and especially on the lady Lilith, the reader may also consult ALLEN, Modern Judaism, pp. 162—168.

of Koenigsberg, that vestiges of the same superstition are found in Sanskrit literature[36].

This belief, that a dead body delights in the blood of a human victim, is likewise met wíth in ancient Greek authors. The phantasm of Achilles is represented, by Euripides, as shewing itself, in golden armour, on his tomb[37], and as appeased by the sacrifice of a young virgin,

[36] I will not attempt to abridge Mr V. BOHLEN's learned and interesting account: "In India are found many demoniacal beings, which may be compared with vampires: among them are *Vetâlâs* (चेताला), evil spirits which dwell in dead bodies, and are also called nightwanderers, flesh-eaters, blood-drinkers, and, by an euphemism, the pure people (punyajanas), somewhat as the Eumenides were in this way propitiated. Farther, to say nothing of the *Rakshasas* (रचसस्) or the *Pisâchâs* (पिंशावा), literally flesh-eaters, who dwell in the deserts, which word Sir WILLIAM JONES (Laws of Menu, I. 37.) translates by *blood-thirsty savages*, but DELONGCHAMPS actually by Vampires. Among the Hebrews the *Lilith*, לילית, ISAIAH, XXXIV. 14. is first to be observed, a female night-spectre, which prostitutes herself to men, and destroys children, resembling the Lamia and Strix, which pursue little boys: and, 'carpere dicuntur lactentia viscera rostris, et plenum poto sanguine guttur habent.' (OVID, Fast. VI. 136.) But the *Aluka* (of PROVERBS, XXX. 15.) comes still nearer as a bloodsucking insatiable monster: the name is derived from עלק (علق in Arabic) and the Kamûs explains the same word العلوقة actually by *algûl*, الغول, the well-known demon of the desert of the Arabian popular stories, which devours men, شيطان ياكل الناس. The notices of vampirism which I find in HORST's Zauberbibliothek are of little importance. A case of vampirism, which occurred at Madvegÿa in Servia, in 1725, became the subject of a judicial investigation, and gave rise to many different publications within the next few years. Among others were STOCKII Dissert. Physica de cadaveribus sanguisugis, Jena 1732. ZOPF, de Vampiris Serviensibus, Halae 1733. POHL, de hominibus post mortem sanguisugis, Lips. 1742. According to Horst there is nothing new in any of them. In the popular songs of Servia, I have only found one passage, which was in my mind during our conversation on this subject: it is in a song, called the Vampire-bride, in the little work LA GUZLA (from which KEIGHTLEY, Fairy Mythology, Vol. II. pp. 322—4. quotes, "the fine Illyrian ballad of Lord Mercury") founded on an event which happened in 1816: but the original is wanting, and one cannot well trust that Guzla. The Servian name of vampire is *Wukodlak*, in Bohemian *Wlkodlak*, that is wolf-hairy; so that at first it must have been considered as a mere incubus."

[37] EURIPIDES, Hec. 109.

whose blood he drank[38]. Oedipus also, in Sophocles, when foretelling a defeat, which the Thebans were to sustain near his tomb, exclaims, that his cold dead body will drink their warm blood[39]. Human victims are offered at the funeral pyre of Patroclus, in the Iliad[40]; and although they are not sacrificed in the Odyssey, yet the blood of the slaughtered sheep is eagerly lapped up by the shades whom Odysseus consults[41]. A sheep used also to be sacrificed at the tombs of mortals, and its blood was supposed to be an offering acceptable to the dead body[42].

38 EURIPIDES, Hec. 533.

Δέξαι χοάς μοι τάσδε κηλητηρίους
νεκρῶν ἀγωγούς· ἐλθὲ δ' ὡς πίῃς μέλαν
κόρης ἀκραιφνὲς αἷμ' ὅ σοι δωρούμεθα.

39 SOPHOCLES, Oed. Col. 621.

Ἵν' οὑμὸς εὕδων καὶ κεκρυμμένος νέκυς
ψυχρός ποτ' αὐτῶν θερμὸν αἷμα πίεται.

40 On the human victims sacrificed by many nations of antiquity, see above, Vol. I. p. 133.

41 ODYSSEY, XI. 45. 48. 95. 96. 153. &c.

42 EURIPIDES, El. 513.

Πυρᾶς δ' ἐπ' αὐτῆς οἶν μελάγχιμον πόκῳ
σφάγιον ἐσεῖδον, αἷμά τ' οὐ πάλαι χυθέν.

Alcest. 860. Θάνατον φυλάξω· καί νιν εὑρήσειν δοκῶ
πίνοντα τύμβου πλησίον προσφαγμάτων.

and v. 866. —— καὶ μὴ μολῇ
πρὸς αἱματηρὸν πέλανον.

All these passages of the tragic writers seem to have escaped from the memory of Bishop BLOMFIELD while he wrote his note on μειλικτήριος; (Gloss. in Aeschyl. Pers. 616.) as well as when he was reviewing Mr Elmsley's edition of the Heraclidae: (QUARTERLY REVIEW, July, 1813. Vol. IX. p. 365.) "The only libations to the dead mentioned by *Greek* authors, consisted of wine, milk, honey, and water. We think therefore that for μήθ' αἷμ' ἐάσῃς should be read μὴ ῥεῦμ' ἐάσῃς." What is, perhaps, still more remarkable, a recent editor of the Alcestis quotes, with approbation, Dr Blomfield's account of these bloodless offerings, on the very passage where the προσφάγματα are mentioned: see MONK, Alc. p. 96. There was even a peculiar word, ἐναγίζειν, appropriated to denote the sacrifice of such προσφάγματα to men below: see VALCKENAER, on Schol. Eurip. Phoeniss. 281. (p. 645.) RUHNKEN, on Tim. Lex. Platon. p. 101. BOISSONADE, on Philostrat. Heroic. p. 518. and KOENIUS, on Gregorius, de dialect. Ion. §. 109. According to the superstition of the Arabs, the Saint Harun (Aaron) is thus propitiated at the present day: see BURCKHARDT, Syria and the Holy Land, pp. 419. 430. Similar offerings are also made to Moses, near the summit of Djebel Músa: BURCKHARDT, l. c. p. 566.

The imaginary beings of many ancient legends resemble still more closely than the phantom of Achilles, the vampire of the present day. At Corinth, Medea's children are said, after their death, to have been in the habit of destroying infants[43]; and only ceased from thus infesting the city, when the Corinthians complied with the injunctions of an oracle, by establishing yearly sacrifices in their honour, and by erecting a statue, in the shape of a hideous woman, over their tomb[44].

According to a story related in the life of Apollonius of Tyana, the long intercourse which took place between a female spectre and the Corinthian Menippus, was a mere prelude to the feast of flesh and blood, in which she meant to revel after their marriage[45].

Strabo[46], Pausanias[47] and other authors[48], record a legend, the scene of which is laid at Temesa in Italy, and which furnishes another example of ancient vampirism. The "demon" of one of Odysseus's companions, who had been stoned to death at Temesa, was believed to infest the place, and equally to attack persons of every age[49]. At length the people prepared to abandon their native land, in order to escape from the calamities

[43] PAUSANIAS, II. p. 118. *Τὰ τέκνα Κορινθίων τὰ νήπια ὑπ' αὐτῶν ἐφθείρετο.*

[44] PAUSANIAS, l. c. and C. O. MUELLER's Aeschylos Eumeniden, p. 141.

[45] PHILOSTRATUS, Life of Apollonius of Tyana, IV. 25. p. 165. *Ἡ χρηστὴ νύμφη μία τῶν Ἐμπουσῶν ἐστιν, ἃς Λαμίας τε καὶ Μορμολυκίας οἱ πολλοὶ ἡγοῦνται· ἐρῶσι δ' αὗται οὐκ ἀφροδισίων μέν, σαρκῶν δέ· καὶ μάλιστα ἀνθρωπείων ἐρῶσιν.* The story ends by the spectre's confession to Apollonius, p. 166. *Ἐμπουσά τε εἶναι, ἔφη, καὶ πιαίνειν ἡδοναῖς τὸν Μένιππον ἐς βρῶσιν τοῦ σώματος· τὰ γὰρ καλὰ τῶν σωμάτων καὶ νέα σιτεῖσθαι ἐνόμιζεν, ἐπειδὴ ἀκραιφνὲς αὐτοῖς τὸ αἷμα.* Since the passage in the text was written, I have found that the story has been already told to the English reader, by Mr GODWIN, Lives of the Necromancers, pp. 159—160.

[46] STRABO, VI. p. 255. [47] PAUSANIAS, VI. p. 467.

[48] AELIAN, V. H. VIII. 18. and SUIDAS, under Εὔθυμος.

[49] PAUSANIAS, l. c. *Τοῦ καταλευσθέντος δὲ ἀνθρώπου τὸν δαίμονα οὐδένα ἀνιέναι καιρὸν ἀποκτείνοντά τε ὁμοίως τοὺς ἐν τῇ Τεμέσῃ, καὶ ἐπεξερχόμενον ἐπὶ πᾶσαν ἡλικίαν.*

which thus befel them; but were told, by the Pythian oracle, that the demon might be propitiated. With this object they were to build a temple, and to offer "to the hero," once a year, the most beautiful virgin of their city. The temple was accordingly raised; access to its sacred inclosure was prohibited to all except the priests[50]; and, thus, it is easy to imagine at what shrine the young and beautiful maiden was annually sacrificed. As soon as a human champion, Euthymus, presented himself on behalf of one of these victims, of lust on the one side and of credulity on the other, the supposed demon fled out of the country[51].

A story, similar to the last in some of its features, is told of a female monster, Sybaris, whose abode was in a cavern on the south side of Mount Parnassus[52]. Eurybatus destroyed the Lamia, and saved Alcyoneus, her destined victim, from his fate.

According to a local superstition of the Lesbians, the unquiet ghost of the virgin Gello used to haunt their island, and was supposed to cause the deaths of young children[53].

We can hardly compare with the vampires of Pausanias and Philostratus, other empusas or spectres, which

[50] AELIAN, l. c. Τὸ ἱερὸν αὐτοῦ, ὅπερ ἄβατον ἦν τοῖς πολλοῖς. MINUCIUS FELIX, Octav. XXIV. p. 225. ed. 1672. Quaedam fana semel anno adire permittunt: quaedam in totum nefas videre. Many examples, of both classes, are mentioned by Prof. LOBECK, Aglaoph. p. 279. The penalty of entering an ἄβατον ἱερὸν was death: ERATOSTHENES, Catast. I.

[51] That the priest acted the demon's part was seen by Perizonius: "ego tamen illum, cum quo pugnavit Euthymus, validum fuisse sacerdotem auguror potius, quam daemonem." An engraving of this "evil genius that infested Temesa," is given by BEAUMONT, in his Treatise of Spirits, and the story is told at p. 18. of the same work. (ed. Lond. 1705.) The engraving and legend are also found in the less common work of CARTARI, Le Immagini degli Dei degli Antichi, and in his French translator, DU VERDIER, Les Images des Dieux des Anciens, p. 531. foll. ed. Lyon. 1581.

[52] ANTONINUS LIBERALIS, c. VIII. Θηρίον μέγα καὶ ὑπερφυές, καὶ αὐτὸ Λαμίαν, οἱ δὲ Σύβαριν ὠνόμαζον.

[53] ZENOBIUS, and HESYCHIUS: MEURSIUS, Lectiones Atticae, III. 17. and ALMELOVEEN, Amoenitates, p. 154.

were objects of terror in many different places; but of whose blood-sucking, flesh-eating or life-destroying propensities nothing is recorded[54]. Thus the ill-fated Io is represented, in the Prometheus of Aeschylus, as pursued by the spectral form of the earth-born Argus[55],

> Who, e'en when dead, remains not in his grave[56].

This phantasm is described as solely discerned by the heated imagination of Io; just as our Shakspeare, in one of the scenes of Hamlet, makes the ghost of the murdered king visible only to his son[57]. The ghost of Actaeon used thus also to terrify the Orchomenians[58]: and we know that the unquiet spectre of Atymnos, the brother of Europa, was believed to haunt the scene of his sister's dishonour, and to appear in the dim twilight, or through the shades of night, to the inhabitants of Gortyna[59].

We also find traces of a principle similar to that in which the modern superstition had its rise, in Plato: he speaks, in the Phaedo, of certain obscure phantasms belonging to impure souls, which had been unable wholly to free themselves from their fleshly prison-house, and which haunted tombs. Plato, however, assigns no such

[54] See TZETZES, on Lycophron, 1030. CONON, c. XXVI. and LOBECK, Aglaophamus, p. 302.

[55] AESCHYLUS, Prom. 567.

> Χρίει τις αὖ με τὰν τάλαιναν οἶστρος,
> εἴδωλον Ἄργου γηγενοῦς.

[56] AESCHYLUS, l. c. 571.

> Ὃν οὐδὲ κατθανόντα γαῖα κεύθει.

[57] When the king's ghost is *visible* it may be compared with the εἴδωλον of Darius, in the Persae, as has been long since observed.

[58] PAUSANIAS, IX. p. 787.

[59] SOLINUS, c. XI. Idem Gortynii et Atymnum colunt, Europae fratrem: ita enim memorant. Videtur hic, et occurrit, sed die jam vesperato augustiori se facie visendum offerens. The dim hour of twilight was naturally that

> When angel forms athwart the solemn dusk
> Tremendous swept, or seem'd to sweep along;
> And voices more than human, through the void
> Deep-sounding, seiz'd th' enthusiastic ear.

On Atymnos see WELCKER, Ueber eine Kretische Kolonie in Theben, p. 8.

blood-sucking propensities to these phantasms, as are attributed to the modern Vampire[60].

It is impossible to leave the subject without regretting our loss of the works of Damascius, among which there was one book "respecting the souls which appear after death." His writings seem, by the admission of the Patriarch Photius, to have possessed considerable literary merit, and would doubtless have made a great addition to what is known of the demonology of the ancients[61].

An unpublished work of Michael Psellus, "on the opinions of the Greeks respecting demons," will probably soon be presented to the Republic of Letters by Mr Hase[62].

In the previous volume[63] I have endeavoured to trace some modern superstitions to their origin, the belief in

Those demons that are found
In fire, air, flood and under ground,
Whose power hath a true consent
With planet, or with *element*.

Other supernatural beings, besides the Holy Virgins and the Vampires, of whom I have heard in Crete, are the Nereídhes[64]: they are said to shew themselves to very few persons, although they constantly frequent the mountains, caves near the sea-shore, and other desert

[60] PLATO, Phaed. §. 69. p. 81. Περὶ τὰ μνήματά τε καὶ τοὺς τάφους κυλινδουμένη, περὶ ἃ δὴ καὶ ὤφθη ἄττα ψυχῶν σκοτοειδῆ φάσματα, οἷα παρέχονται αἱ τοιαῦται ψυχαὶ εἴδωλα, αἱ μὴ καθαρῶς ἀπολυθεῖσαι, ἀλλὰ τοῦ ὁρατοῦ μετέχουσαι. The same notion was prevalent among the Jews: see LIGHTFOOT, on JOHN, XI. 39. ELSNER, Obs. Sacr. p. 47. (compare MATTHEW, VIII. 28. MARK, V. 1. LUKE, VIII. 26.)

[61] On Damascius and his writings, see PHOTIUS, Biblioth. cod. 130. We might perhaps have found more details, respecting the prototype of the modern Katakhanás, in the work, Περὶ τῶν μετὰ θάνατον ἐπιφαινομένων ψυχῶν παραδόξων διηγημάτων κεφάλαια ξγ'.

[62] HASE, on Leo Diaconus, Lib. VI. p. 464. "Michaelis Pselli libellus Graece nondum editus de opinionibus Graec. circa daemones, (quem volumini proximo historiae eiusdem Pselli fortasse subiiciam.)"

[63] Vol. I. pp. 89—93.

[64] Νερηΐδες.

spots: they are of the female sex, are very beautiful, and always pass their nights in dancing[65].

Thus it appears that the Nereid nymphs of antiquity[66] have preserved their sex, and personal beauty, and even their ancient attachment to the dance[67], along with their watery habitations in dripping caverns near the sea. The title of an unpublished treatise, contained among the manuscripts of the Vienna Library is, "concerning the nymphs whom the common people call Nereids[68]." Another epithet, "the beautiful ladies,"

[65] A modern traveller in Crete when at Goniés, on the north-eastern slopes of Mount Ida, seems, although wholly ignorant of the language of the island, to have found some slight traces of this belief: see SIEBER, Reise, Vol. I. p. 432.

[66] To whom a sacred inclosure was dedicated, near Cardamyle in Laconia: PAUSANIAS, III. p. 278. and who were represented on several works of ancient sculpture: PAUSANIAS, II. p. 112. and V. p. 426. and are also frequently found on coins and gems, MUELLER, Handbuch der Archaeologie, 402. 3. (p. 615. 2d. ed.)

[67] EURIPIDES, Ion, v. 1081.

> Καὶ πεντήκοντα κόραι
> Νηρέος, αἱ κατὰ πόντου
> ἀεννάων τε ποταμῶν
> δίνας χορευόμεναι—

Compare Iphig. Taur. 427. Iphig. Aul. 1057. and AESCHYLUS, in the Scholia on Aristophanes, Acharn. 848. Their number varied: see PLATO, Critias, Tom. III. p. 116. e. and other passages indicated by MUSGRAVE, on Soph. Oed. C. 718, 9. It is hardly worth while to stop to consider, in this place, whether the daughters of Nereus ought to be distinguished from the sea-goddesses the Nereids: see EUSTATHIUS, p. 1954, 5. AMMONIUS, p. 97. and Valcken. Animadv. p. 161. It is deserving of notice that the proper form of the name, Νηρεΐδες, does not occur in any of the passages where they are mentioned by Attic writers: see F. ELLENDT, Lex. Sophocl. Vol. II. p. 160.

[68] NESSEL, Catalog. Biblioth. Vindob. P. V. p. 168. L. N. A'. The work is by Joannes Magister Canabutius, and is entitled Περὶ Νυμφῶν, τίνες εἰσί, καὶ πόσα γένη τούτων εἰσί, καὶ ὅτι γένη δαιμόνων εἰσίν, ἃ καλοῦσι βαρβαρίζοντες Νεραγίδας, quae vulgo vocantur Νεραΐδες: the γ may be inserted or not, as in many other modern Greek words, and makes hardly any perceptible difference in the sound. Thus παναγία has given way to παναΐα, in the text of Hesychius: see MEINEKE Quaest. Scen. II. 39. WELCKER, Sylloge Epigramm. Graec. p. 121. and thus also we have κλαίει and κλαίγει &c. &c. and the common word ἅγιος has nearly the same sound as ἅϊος. This title is quoted by Professor LOBECK, Aglaoph. p. 1204. KORAY, ΑΤΑΚΤΑ, Tom. II. p. 257. says, Νεράδες, θεαὶ θαλάσσιαι, τὰς ὁποίας

bestowed on these creatures of popular superstition[69], reminds us of the Eumenides of the ancients, and of other euphemisms used to propitiate the imaginary beings of our own fairy mythology.

The lofty mountains are the favourite haunt of the Nereids at the present day. Even among the ancients the same transfer of Water-Nymphs to the mountains was common. The huntress and mountain-goddess Artemis is attended, as we have seen, by Cretan River-Nymphs[70]; and Cyrene, in Virgil, invokes at once

> Oceanumque patrem rerum, Nymphasque sorores,
> Centum quae silvas, centum quae flumina servant.

These modern Nereids bear a closer resemblance to the fairies of Old England, than any other supernatural beings, of whom I have heard in the East. Our fairies principally inhabit mountains and caverns, and their chief nightly pastime is dancing[71].

The main intercourse of "the beautiful ladies" with mankind, is when they change one of their young Nereids[72] for the child of a woman, who has just become a mother. Their offspring, so left in exchange, is said never to live. The conduct of the Cretan Nereid re-

ὁποίας πιστεύει ἀκόμη ὁ ὄχλος, χωρὶς ὅμως νὰ γνωρίζῃ καλὰ ποῦ κατοικοῦν: the Cretans, however, think they know this full well, and told me, as I have mentioned in the text.

[69] KORAY, ATAKTA. Tom. 4. p. 211. *Πιθανὸν ὅτι εἶναι κατ' εὐφημισμόν, ὡς καὶ αἱ* Νηρηΐδες, ʽΕλλ. *ὀνομάζονται κοινῶς, ὄχι μόνον* Νεραΐδες (Néréides), *ἀλλὰ καὶ* Καλαὶ *ἀρχόντισσαι* (belles dames).

[70] Above, Vol. I. p. 266. and SPANHEIM, on Callimachus, Hymn to Artemis, v. 13. HEYNE, on Virgil, Georg. IV. v. 343.

[71] KEIGHTLEY, Fairy Mythology, Vol. II. Chapter on Great Britain, p. 104. CROKER, Fairy Legends and Traditions of the South of Ireland, Part III. p. 90. "Wherever the fairies hold a feast, they are accompanied by music: nor is it wanting in their large and festive processions: in this the traditions of all nations are unanimous." The same abode in the mountains, and fondness for dance and song, also characterize the Servian *Vila:* and all the mountain and water spirits of the popular belief of Northern Europe, equally delight in music and dancing: see J. GRIMM, Deutsche Mythologie, pp. 264. and 278.

[72] Ἕνα Νεραϊδάκι.

sembles, in this point also, the general practice of our fairy[73].

When larks 'gin sing
Away we fling,
And babes new born steal as we go,
And elfe in bed
We leave instead,
And wend us laughing: Ho! Ho! Ho![74].

I, however, find no trace, in Crete, of the belief in the diminutive stature of the Nereids, a peculiarity of our English fairies, which they have been supposed to have derived, with some of their unamiable qualities, from the Gothic elves[75].

The following Sfakian story[76] also has a resemblance to the legends of our northern mythology, with its

Fairy elves,
Whose midnight revels by a forest side,
Or fountain, some belated peasant sees,
Or dreams he sees, while over head the Moon
Sits arbitress—they on their mirth and dance
Intent, with jocund music charm his ear[77].

"Two men went, on a fine moonlight night, up the lofty mountains, intending to hunt the agrímia. They heard a great tumult, and at first supposed it to be caused by people coming to obtain snow, to take into the city: but, as they drew nearer, they heard the

[73] SCOTT, On the Fairies of popular superstition, in Vol. II. p. 167. foll. of the Minstrelsy of the Scottish Border: CROKER'S Researches in the South of Ireland, Chapter V.: and CROKER, Fairy Legends and Traditions, Vol. I. pp. 77—80. The belief is alluded to in SHAKSPEARE, Henry IV. Part I. Act I. Sc. I. and SPENSER, Faerie Queene, I. 10. 35. See BRAND'S Popular Antiquities, Vol. II. p. 8. foll.

[74] See also DRAYTON, Nymphidia, Stanza X. The same opinion was entertained of an evil spirit as early as the time of MICHAEL PSELLUS: see his work de operat. daem. p. 78. (and WARTON, on Milton's L'Allegro.)

[75] SCOTT, Minstrelsy of the Scottish Border, Vol. II. p. 178. No notice of this diminutive stature is found in Chaucer's account of them: see KEIGHTLEY, Fairy Mythology, Vol. II. p. 121.

[76] I wrote down the story in the very words of my Sfakian informant, as it is given in Note C, at the end of the Chapter. The text contains a condensed translation of it.

[77] MILTON, Paradise Lost, I. 781.

sound of musical instruments and varied sports. The men soon discovered that these were not mortals, but an assemblage of demoniacal beings[78]: all of whom were clothed in varied garments, and rode on horses, some of which were white, and others of different colours. It appeared that there were "both men and women, on foot and on horseback, a multitude of people: and the men were white as doves, and the women beautiful as the sunbeams:" it was also evident that they were carrying something like a bier. The mountaineers determined to shoot at the aerial host, as they passed on singing,

> We go, we go, to fetch the lady-bride
> From the steep rock, a solitary nymph.

As soon as the shot was fired, those who were last in the procession exclaimed, 'They've murdered our bridegroom—they've murdered our bridegroom:' and, as they made this exclamation, they wept, and shrieked, and fled."

Some passages in an account of the *fairy fowks rade*, shew an identity, both of fact and phrase, so curious as to make it well worth while to compare the whole story of the Cretan mountaineers, with that of an old woman of Nithsdale, as given in Mr Keightley's work[79].

As the idea that the unearthly beings were mere mortals, going to fetch snow, struck the Sfakians, so the Scot at first thought that "it was drunken fowk riding to the fair in the forenight." As the Sfakians had never before heard such sounds, so a young North-Briton when out, on a love affair, heard most delicious music, far surpassing the utterance of any mortal, "and advancing to the spot whence the sound appeared to proceed, suddenly found himself the spectator of a fairy-banquet[80]."

[78] The words were *δαιμονικὸ συνέδριον*.
[79] KEIGHTLEY, Fairy Mythology, Vol. II. p. 166.
[80] See KEIGHTLEY, l. c. p. 162.

My hostess here, at Anópolis, was once traversing the mountains, accompanied by one of her daughters, and, when about three miles from the village of Murí, they heard sounds as of voices singing[81], but it was impossible to distinguish what were the words uttered[82]. The demon, for such she supposed the unseen object of her alarm to be, then began to throw stones, which fell both before and behind them. Although she saw him not, yet she immediately pronounced aloud some holy texts, which are a never failing charm against any common demon[83]. When she found that the evil spirit continued to sing, and to cast stones at them, she knew that it must be a Katakhanás; and, therefore, crossing herself, and calling on the Holy Mother of God, she immediately repeated: "In the beginning was the Word, and the Word was with God, and the Word was God[84]." This sentence she pronounced thrice, but it was all in vain: the Katakhanás kept rolling down the stones[85] as before. She next repeated a part of the Greek ritual[86], which produced no better effect: the Katakhanás continued to persecute and terrify her. At length, on seeing two women, as she approached the village, she summoned courage enough to address him in a bolder strain[87], and, on her doing so, he ceased to molest her.

An ancient Cretan, under the same circumstances, would probably have called on as many names as were invoked by the old lady. Instead of the Panaghía, the Lógos, and the Angels, he would have used the celebrated Ephesian words[88], or the names of the Idaean

81 *'Ωσὰν τραγοῦδι.* 82 *"Ομως δὲν ἐκαθαρίσθη τὶ εἶπεν.* 83 *Δαίμονας.*

84 *'Εν ἀρχῇ ἦν ὁ ῥύγος, καὶ ὁ ῥόγος ἦν πρὸς τὸν θεόν, καὶ θεὸς ἦν ὁ ῥόγος.* 85 *'Εῤῥήκτει τζὴ πέτραις.*

86 She said it from the *ἦχος* of the Panaghía, and began thus: *Ἄγγελος πρωτοστάτης οὐρανοῦ*, &c.

87 Her words were, *'Εξελθέ, διάβολε, ἀπὸ μᾶς, διατὶ θὰ σοῦ βάῤῥω τὸ ῥαβδί μου εἰς τὸ κώλο σου*: as soon as she said this, *ἔγενε ἄφαντος.*

88 EUSTATHIUS, on the Odyss. p. 1864, 14. *'Εφέσια γράμματα, ἐπῳδαί τινές, φασιν, ἐκεῖνα ἦσαν.* See CLEMENS ALEX. Strom. v. p. 568. and Potter's note, HESYCHIUS, in v. *'Εφέσια γράμματα*: WETSTEIN, on Acts

Dactyls, who were regarded as beings of a higher order, and powerful enough to avert, whenever they were invoked, any impending calamity from the race of men[89].

Similar malevolent demons, which used to molest men, and still more women, even in the daytime, existed in the superstition of ancient times. A Jewess, if well read in the Septuagint version of her Scriptures, would probably have taken the supposed supernatural being for "the mid-day demon," which is spoken of in the Psalms[90]: and a heathen ancestress of my hostess would have believed the persecuting phantasm to have been sent by Dictynna or Artemis[91].

The author of a book commonly attributed to Hippocrates[92], treats the belief in similar visions of hostile demons as a disease. He says, that such imaginary phantasms, supposed to be seen not only by night but by day, had caused many persons, especially women, to hang themselves; and subjoins, that on recovery from such a disorder, it was usual for the women to make expensive offerings to Artemis, at the instigation of the priests, who, as usual, profited by their credulity[93].

Acts, XIX. 19. MEINEKE, on Menander, p. 132. and, most of all, WYTTENBACH, on Plutarch, Tom. VI. Pars I. p. 605. and LOBECK, Aglaophamus, pp. 1163. and 1330—1333.

[89] PLUTARCH, de Sent. Profect. in Virtut. p. 85. 3. (p. 316. Reisk.) *Οἱ μὲν γὰρ ἐκμεμαθηκότες τὰ τῶν Ἰδαίων ὀνόματα Δακτύλων χρῶνται πρὸς τοὺς φόβους αὐτοῖς ὡς ἀλεξικάκοις, ἀτρέμα καταλέγοντες ἕκαστον.*

[90] PSALMS, XCI. 6. *Οὐ φοβηθήσῃ—ἀπὸ πράγματος διαπορευομένου ἐν σκότει, ἀπὸ|συμπτώματος καὶ δαιμονίου μεσημβρινοῦ.* See LE CLERC, Ars Critica, P. I. c. I. p. 52. foll.

[91] See LOBECK, Aglaophamus, p. 1092. who cites the Acta Symphoriani Ruinart. Act. Mart. p. 70. "Dianam *daemonium esse meridianum* sanctorum industria investigavit, quae per compita currens, *et silvarum secreta perlustrans,* hominum mentibus zizaniae tribulos disseminat."

[92] See KUEHN, Hist. Literar. Hippocrat. p. CXXXIV. 20.

[93] HIPPOCRATES, de Morb. Virg. p. 357. or Tom. II. p. 526. ed Kuehn. *Περὶ τῶν δειμάτων ὁκόσα φοβεῦνται οἱ ἄνθρωποι ἰσχυρῶς, ὥστε παραφρονέειν καὶ ὁρῆν δοκέειν δαίμονας τινὰς ἐφ' ἑωυτῶν δυσμενέας, ὁκότε μὲν νυκτός, ὁκότε δὲ ἡμέρης—ἔπειτα ἀπὸ τῆς τοιαύτης ὄψιος πολλοὶ ἤδη ἀπηγχονίσθησαν, πλέονες δὲ γυναῖκες ἢ ἄνδρες.* Again, p. 528. *Φρονησάσης δὲ τῆς ἀνθρώπου, τῇ Ἀρτέμιδι αἱ γυναῖκες ἄλλα τε πολλά, ἀλλὰ δὴ καὶ τὰ πολυτελέστατα τῶν ἱματίων καθιεροῦσι τῶν γυναικείων,*

κελευόντων

Since the mid-day demon has thus brought us back to the Katakhanás, I will notice one other class of actions sometimes performed by that fearful being, and which are of a very different character from those already described. When he visits a wife, in the absence of her husband, she usually survives the interview[94]. Thus, as in other ages and countries, a mortal lover has at times been converted into a god[95], an angel[96], or a saint; so, in Crete, he has, doubtless not unfrequently, been believed to be a vampire.

During the long war between the Christians and Mohammedans of the island it became a matter of astonishment, that ravages caused by vampires were no

κελευόντων τῶν μάντεων, ἐξαπατεώμεναι. With this mention of Diana, Professor LOBECK, l. c. compares the words of HORACE, A. P. 453.

94 Ἕνας Καταχανᾶς ἐγύρισε εἰς τὴν Ἀνώπολιν, καὶ ἐπλάκωνε τζὴ ἀνθρώπους, καὶ ἐγάστρωσε καὶ μίαν γυναῖκα. Ὁ ἄνδρας τζή εἶχε λείπει καὶ ἐπῆγεν ἕνας καταχανᾶς, καὶ αὐτὴ θάρρωντας ὅπως εἶναι ὁ ἄνδρας τζή—καί, τὸ πουρνό, δὲν ἠμπόρει· καὶ ἦλθε καὶ ὁ ἄνδρας τζὴ, καὶ λέγει τζή, "τί ἔχεις;" καὶ ἡ γυναῖκα λέγει "πολλὴν ὥραν με ἐπλάκωσες τὸ βράδυ, καὶ δὲν ἐμπορῶ·" καὶ ὁ ἄνδρας λέγει, "ἐγὼ δὲν ἦλθα." καὶ πάλιν ἡ γυναῖκα εἶπε, "ἐδά ὡσὰ δὲν ἦλθες ἐσύ, ἤτονι ὁ Καταχανᾶς," an explanation with which the husband was fully satisfied.—ὑστερινὰ τὸν ἐξεχώσασι, καὶ ἐξορκίσανι τόνι καὶ ἔπεμψαν τὸν εἰς τὴν Σαντορίνην. With the performance of this Katakhanás the exploit of BOCCACCIO's pallafreniere (Decamerone, Giornata III. Novella II.) deserves to be compared. Of the truth of the Sfakian story, and that the individual was a genuine Katakhanás, both my host and his wife were fully satisfied: compare WIERUS, de Praestigiis Daemonum, Basil, 1564. Lib. II. c. 39. and other parts. The title of c. 39. is "Incubi illusionem infestare quandoque probas etiam matronas, cum ridiculo exemplo adulterini congressus daemoniaci."

95 See BAYLE, Dictionnaire Philosophique, v. Scamandre; WOOD, Essay on the original genius of Homer, p. 2. GIBBON, end of c. XXVIII. JOSEPHUS, A. J. XVIII. 3, 4. p. 799. Paulina sleeps, καὶ ὁ Μοῦνδος, προεκέκρυπτο γὰρ τῇδε, οὐχ ἡμάρτανεν ὁμιλιῶν τῶν πρὸς αὐτήν· παννύχιόν τε αὐτῷ διεκονήσατο, ὑπειληφυῖα θεὸν εἶναι. καὶ—πρωῒ ὡς τὸν ἄνδρα ἐλθοῦσα, τὴν ἐπιφάνειαν ἐκδιηγεῖται τοῦ Ἀνούβιδος, just as the Sfakian lady told her husband of the Katakhanás.

96 I need hardly name Boccaccio as an ample authority for this assertion. We are even told by grave Roman Catholics, to use the words of an English bishop, "that 'once Christ came, in company with St Dominic, to visit Teresa: Christ soon withdrew, and bade her recreate herself with his friend Dominic, who stayed with her two hours.' Such is the language and effect of *spiritual love* among the Popish fanatics." Dr LAVINGTON, (Bishop of Exeter,) The Enthusiasm of Methodists and Papists compared, Vol. I. p. 54. 3d. ed.

longer spoken of. How can it be when the number of deaths is so great, that none of those who die become Katakhanádhes? The answer given was, "No one ever becomes a Katakhanás[97], if he dies in time of war"!

Another superstitious belief, universally prevalent in the island, is, that if a Papás excommunicates, or even utters an imprecation against, any one[98], the man falls ill, and can only be restored to health by going to the very Papás who has caused his suffering, or to a bishop, to obtain the removal of the ban. If a Therisióte, Lakióte, or Sfakian, has a head-ache, fever, or rheumatism, it is at once the excommunication of a Papás which has done it[99].

Thus the clergy, availing themselves, to the fullest possible extent, of the ignorance and credulity of the people, have extended the bounds of their long recognized dominion over the soul, even to the body: and, more than this, not content with the power of rendering any one a restless vampire after death, they convince their flocks that the enjoyment even of bodily health, during life-time, wholly depends on the will of their spiritual rulers and guides!

The influence which such a popular superstition bestows on the clergy is sufficiently obvious: and the sincerity of the belief of these rude mountaineers, in the power of their priests over the body as well as the soul, has at times led to disagreeable, and, more than once, even to fatal results.

97 *Δὲν καταχανεύουνι.*

98 I was told that his calling a man *θεοκατάρατος* would suffice to produce results similar to those of the regular *ἀφορισμός*.

99 This superstition was equally general in the sixteenth century. The Proveditor FOSCARINI, (in his Relatione, &c. foll. 115. date 1576.) says that the Cretans "alli loro papati portano grandissimo rispetto et riverentia, massime li contadini et gente bassa, che temono più la scomunica del Papá che non fanno tutto il male che potesse loro avenire: et se per accidente li occorre qualche disconzo, o di malattia o d'altro, quando possono dubitar di essere stati escomunicati, si riducono in disperatione, ne possono potersi mai liberar da quel male, se non sono dal medesimo Papá, con modi superstitiosi et quasi idolatri, rebenedetti."

Some years ago a Greek of Kerámia complained to the Pashá of Khaniá, that the Papás of his village had thus excommunicated or bewitched him; whereon the Pashá sent for the priest, threw him into prison, and only released him on his paying a fine of three hundred piastres.

During the war, a native of Thériso was taken ill: it is an *ἀφορισμός*, was the general cry: the Papás was reviled and threatened: still the man grew worse, and at last died. So firm was the belief of every one in the neighbourhood, that the ban had caused his death, that a few of his companions felt it a duty to avenge his fate, and immediately sought out and shot the poor Papás.

A story, which also serves in a still more striking manner to exemplify the crimes caused by these superstitious notions, is told of a bishop of some diocese in the Moréa, who was once robbed as he was passing through a part of the Maniáte territory. When the deed had been performed, the mountaineers called to mind that the bishop would, in all probability, excommunicate them, as soon as he reached a place of safety. They saw no means of averting this dreadful calamity, except by the death of the unfortunate prelate, whom they therefore pursued, overtook and shot.

Whether we sigh or smile at the narrative of such credulity and superstition, we must remember, that far greater atrocities were committed, as late as the sixteenth and seventeenth centuries, in our own country; where they were even invested with the sanction of acts of parliament, and were uniformly approved by the clergy. In the age of Shakspeare, a pious bishop believed that he could see most evident and manifest marks of wickedness, of which the victims pined away even unto death: "Their colour fadeth," says Jewel, in a sermon preached before Elizabeth, "their flesh rotteth; their speech is benumbed; their senses bereft:" and, in the first year of her successor's reign, an act of parliament was passed, which appointed "the pains of death" as the punishment of enchantments and witchcraft. Even Sir Paul Ricaut, when writing, near the end of the seventeenth century, of similar Greek superstitions, adds, that the instances mentioned were told "with as much variety as we do the tales of witches and enchantments, of which it is observed in conversation, that scarce one story is ended before another begins of like wonder[100]."

In Great Britain it is undoubted that not only hundreds but thousands of unhappy victims, of both sexes, suffered a cruel death for this imaginary crime. Unfortunately for the interests of humanity, the clergy were not supposed, with us, to be in exclusive possession of the art of witchcraft, as they are now in Greece: and, since they generally shared in the superstition of the day, they contributed, in no slight degree, to increase the popular violence and cruelty. For the imaginary power which, if wielded, as it is in Greece, by those who are expressly set apart for the service of heaven, would have been regarded as derived from God, became, when in lay hands, a gift of the Devil: and the well-known text, "thou shalt not suffer a witch to live," was of course, considered, in that age, both as a proof of the reality of the crime, and an incitement to exterminate all those whom the miserable wretches called witch-finders might accuse of having committed it[101]. The royal demonologist, James, expressly lays down,

[100] RICAUT, The present State of the Greek Church, p. 278.

[101] In Scotland, especially, the superstition of the clergy made them encourage the witch prosecutions, and approve of the invariable torture by which they were accompanied, and of the death by burning, in which they ordinarily ended: see Sir WALTER SCOTT, Letters on Demonology and Witchcraft, Letter IX. p. 304. foll.: "These venerable persons entertained, with good faith, the general erroneous belief respecting witchcraft, regarding it indeed as a crime which affected their own order more nearly than others in the state, since, especially called to the service of Heaven, they were peculiarly bound to oppose the incursions of Satan." Sir Walter Scott observes, at p. 317, that the English sectarians in Cromwell's time beheld with horror and disgust "a practice so inconsistent with their own humane principles of universal toleration."

that "never so diffamed persons may serue for sufficient witnesses and proofes in such trialls:" and, as late as the year 1664, Sir Matthew Hale told a jury, that since the Scripture left no doubt that there was such a thing as witchcraft, "the only question to be considered by them was, 1st, whether certain children were really bewitched, and 2d, whether the witchcraft was sufficiently brought home to the prisoners at the bar;" two unfortunate women, who were found guilty by the jury, were sentenced by the judge[102], and were duly hanged by the executioner[103].

102 "In conclusion the judge and all the court were fully satisfied with the verdict, and thereupon gave judgement against the witches that they should be hanged. They were much urged to confess, but would not. That morning we departed for Cambridge, but no reprieve was granted: and they were executed on Monday the seventeenth of March following, but they confessed nothing." A Tryall of Witches at the Assizes held at Bury St Edmond's, for the County of Suffolk; on the tenth day of March, 1664, before Sir Matthew Hale, Kt. &c." London, 1682. pp. 58–59.

103 The last death in Scotland, for the same crime, was in the year 1722. Sir WALTER SCOTT, Demonology, Letter IX. p. 338. The statute of James was not repealed till just a hundred years ago (in 1736). In France Louis XIV. had forbidden all judicial investigation into alleged cases of sorcery more than half a century earlier, and only eight years after Sir Matthew Hale had pronounced his sentence of death on a similar charge. See VOLTAIRE, Siècle de Louis XIV. Ch. XXXI. GODWIN, Lives of the Necromancers, p. 464.

ADDITIONAL NOTES

TO

CHAPTER XXXVI.

NOTE A. (p. 197.)

ΕΙΣ ἕνα καιρὸν παλαιὸν στὸ χωριό Καλικράτι, τῆς ἐπαρχίας Σφακιά, ἐβγῆκεν ἕνας Καταχανᾶς, καὶ δὲν ἐξεύρασι τὶ ἄνθρωπος ἦτονι, ἢ ἀπὸ τὶ μέρος. Τοῦτος ὁ Καταχανᾶς ἐχάρασε παιδιά καὶ ἀνθρώπους μεγάρους πoῤῥούς, καὶ ἐρήμνωξε καὶ τὸ χωριό ἐκεῖνο καὶ ἄῤῥα ποῤῥά. Καὶ τὸν εἶχασι θαμμένον εἰς τὸν ἅγιον Γεώργιον, εἰς τὸ Καλικράτι, καὶ κατὰ καιρὸν ἐκεῖνον ἦτον σημαντικὸς ἄνθρωπος, καὶ τοῦ εἴχασι καμάραν καμομένην. Ἦτονι καὶ ἕνας βοσκὸς σύντεκνος τοῦ μεροτικός, καὶ ἔβοσκε ὡζὰ ἐκεῖ, κοντὰ εἰς τὴν ἐκκλησίαν· ὅμως τὸν ἐπίασε βροχή, καὶ ἐπῆγε στὸ μνῆμα, νὰ σκιασθῇ ἀπὸ τὴν βροχή. Μετ' αὐτὸ ἀποφάσισε διὰ νὰ κοιμηθῇ, καὶ νὰ ξωμείνῃ ἐκεῖ· καὶ βγάνει τὰ ἅρματά του, καὶ τὰ θέτει στὸ προσκεφάλι του, σταυροτά· (καὶ διὰ τοῦτο εἶχε νὰ 'πῇ ὁ κόσμος, ὅπως διὰ τοῦτο δὲν εἶχε τὴν ἄδειαν νὰ πορίσῃ ὁ Καταχανᾶς·) Τὴν νύκτα, λειπόν, αὐτός, ὅπως ἠθέλησε νὰ πορίσῃ πάλιν, διὰ νὰ πάγῃ νὰ χαράσῃ ἀνθρώπους, λέγει τοῦ βοσκοῦ· "Σήκωσε, σύντεκνε, ἀπ' αὐτοῦ, ἐπειδὴ ἔχω δουλιά διὰ νὰ πορίσω. Αὐτὸς δὲν τοῦ ἀποκρίθηκε, μήτε στὴν μία φοράν, μήτε τζὴ δύω, μήτε τζὴ τρεῖς· τοιούτως ἐγνώρισεν αὐτὸς ὅπως εἶναι καὶ αὐτὸς καταχανημένος, καὶ εἶναι αὐτὸς ἀποῦ ἔκαμεν ἐκεῖνα τὰ κακά. Διὰ τοῦτο τοῦ ὡμίλησεν, τζὴ τέσσεραις ἀπάνω, καὶ τοῦ εἶπε· "Δὲν σηκώνομαι ἀπ' ἐδά, διατὶ φοβοῦμαι, σύντεκνε, μὴ μπὰ νὰ ἔχῃς ἀπὸ κακὸν αἱέρα πρᾶγμα παθομένο, καὶ με πειράξῃς, καὶ φοβοῦμαι· ὅμως, σὰ θέλεις νὰ σηκωθῶ, μνώξέ μου 'μὰ τὸ ἀναβόλι σου,' ὅπως δὲν με πειράζεις, καὶ ἔτσι σηκώνομαι." Καὶ αὐτὸς δὲν τοῦ τὸ ἔλεγε, παρὰ τοῦ ἔλεγεν ἄῤῥα· ὅμως σὰ δὲν τὸν ἄφινε *νὰ σηκωθῇ, τοῦ ἔμνωξε ὡς καθῶς 'ποῦ ἤθελε· καὶ σηκώνεται ἀπ' ἐκεῖ, καὶ πέρνει τὰ ἅρματά του, καὶ τὰ βγάνει ἀπὸ τὸ μνῆμα ὄξω.* Καὶ *αὐτὸς, (ὁ καταχανᾶς,) ἐπόρισε, καὶ τὸν ἐχαιρετιχτήκασε, καὶ τοῦ λέγει,* "Ὅμως *σύντεκνε, μὴ μισέψῃς, ἀῤῥὰ κάθισε ἐδὰ, ἀποῦ ἔχω δουλιά διὰ νὰ πάγω, καὶ εἰς μίαν ὥραν θὲ νὰ γυρίσω· ἀποῦ ἔχω λόγον νὰ σου 'μιλήσω.* Καὶ *αὐτὸς τὸν ἐκαρτέρησε.*

Καὶ *ὁ* Καταχανᾶς *ἐπῆγεν ὡς δέκα μίλια δρόμου, ἀποῦ ἤτονι ἕνα ἀνδρόγυνον εὐλογισμένον, καὶ τοὺς ἐχάρασε καὶ γιαίρει ὀπίσω, καὶ τὸν εἶδεν ὁ σύντεκνός του ἀποῦ ἐβάστα τὸ σκότι, καὶ ἤτανι τὰ χέριά του ὀγρὰ ἀπὸ τὸ αἷμα, καὶ ἐκεῖ ἀποῦ τὸ ἐβάστα τὸ ἐφύσηξεν, ὡς καθῶς ἀποῦ φυσᾷ ὁ χασάπις διὰ νὰ μεγαλώσῃ τὸ σκότι· καὶ τοῦ ἔδειξεν ὅπως ἦτον ψημμένον, ὡς καθῶς καὶ εἰς τὴν φωτιάν.* Μετ' *αὐτὰ τοῦ εἶπε, νὰ καθίσωμεν σύντεκνε, νὰ φάμε· καὶ αὐτός ἔκαμεν ὅπως ἔφαγε, ὅμως ἔτρωγε ξηρὸν ψωμί, καὶ ἀπὸ ἐκεῖνο δὲν ἔφαγε, ἀῤῥὰ τὸ ἔβαλλε στὸν κόρφον του.* Διαταῦτος *ἦλθεν ὥρα ὅπως νὰ ἀναχωρήσωσι, καὶ τοῦ εἶπεν ὅπως,* "Σύντεκνε, *τοῦτο ἀποῦ εἶδες, νὰ μὴ τὸ μαρτυρήσῃς, διατὶ τὰ εἴκοσι μου ἀνύχια εἰς τὰ παιδιά καὶ εἰς τὸν ἀπατόν σου θὰ εἶναι κρεμασμένα."* Καὶ *αὐτὸς ὅμως δὲν ἔχασε καιρόν, παρὰ ἔδωκεν τὴν εἴδησιν εἰς ἱερεῖς, καὶ εἰς ἐπιλοίπους ἀνθρώπους, καὶ ἐπήγασαν καὶ εὑρήκασάν τον εἰς τὸ μνῆμα, καὶ ἐγνωρίσασιν ὅροι ὅπως αὐτὸς εἶχε ὅρα τὰ κακὰ καμομένα· διαταῦτος ἐμαζέψασιν οἱ ἄνθρωποι ξύρα ποῤῥά, καὶ τὸν ἐβάλασιν ἀπάνω, καὶ τὸν ἐκάψασιν· καὶ ὁ σύντεκνός του δὲν ἦτον εἰς τὸ εὔγορον, καὶ ὅταν ἤτονι μεσοκαιεμένος ἐπρόβαλε καὶ αὐτὸς διὰ νὰ κάμῃ χάσι, καὶ ἐπέταξεν ὁ* Καταχανᾶς *ὡς καθῶς μίαν ῥουχάλα μὲ αἷμα, καὶ τοῦ ἔδωκεν εἰς τὸ πόδι του, καὶ ἐλύθηκεν ὡς νὰ ἤτονι εἰς τὴν φωτιάν ψημμένον· διαταῦτος ἐκοσκινίσασι καὶ τὸν ἄθον, καὶ εὑρήκασι τὸ μικρὸ ἀνυχάκι τοῦ* Καταχανᾶ *ἄκαγον, καὶ τὸ ἔκαψαν καὶ ἐκεῖνο.*

NOTE B.

Leo Allatius, in his Essay de Quorundam Graecorum Opinationibus, devotes many pages (pp. 142–158) to this subject. I extract the following passages:

P. 142. "Et haec ferenda videntur, si unum excipias Burculacam; alii Bulcolaccam, alii Buthrolacam vocant; quo sane in genus humanum nihil potest excogitari immanius aut perniciosius. Nomen est inditum a foeditate. *Βοῦρκα* limus est, non quilibet, sed qui jam putrescenti aqua maceratus, pessimam exhalat Mephitim, ut ita dicam. *Λάκκος* fossa, seu cavea, in qua similis limus fovetur. Est porro pessimi hominis, et facinorosi, saepeque etiam ab Antistite suo excommunicati cadaver, quod, non ut reliqua demortuorum corpora defossa dissolvuntur atque in pulverem abeunt, sed quasi ex firmissima pelle constaret, per omnes sui partes intumescit, atque distenditur, ut vix flecti aliqua sui parte possit; sed cutis, tanquam tympanum, extensa, eundem, ac tympanum, si pulsatur, sonum edit; quare et *τυμπανιαῖος* dicitur. Corpus, sic deformatum, Daemon ingreditur, et miseris mortalibus infortunium parit. Saepe enim sub eo cadavere e sepulchro egressus, et per urbem et alia loca habitata circumiens, et noctu potissimum, ad quam sibi libuerit aedem, confertur, pulsatisque foribus, aliquem ex accolis aedis voce sonora compellat. Si responderit, actum jam est de eo: altero enim die mortem obit. Si non responderit, salvus est. Hinc in ea insula cives omnes, si noctu ab aliquo compellantur, nunquam prima vice respondent: nam, si secundo compellatus fuerit, jam, qui quaerit, Burcolacca non est, sed alius. Eamque pestem adeo exitiosam mortalibus esse dicunt, ut interdiu etiam, et meridie ipso, non intra aedes tantum, sed in agris, et viis mediis, et sepibus vinearum, praetereuntes aggrediatur, et aspectu solo ac visu conficiat, non verbis tantummodo et contactu enecet. Homines ipsi, qui viderunt, si alloquantur, spectrum disparet; qui locutus est, moritur. Quare cives, cum vident homines, nulla grassante infirmitate, in tanta copia emori; suspicati quod est, sepulchra, in quibus recens defunctus sepultus est, aperiunt; aliquando statim, aliquando etiam tardius, cadaver nondum corruptum, inflatumque, comperiunt; quod e sepulchro extractum, precibusque effusis a sacerdotibus, in rogum ardentem conjiciunt; et nondum completa supplicatione, cadaveris juncturae sensim dissolvuntur, et reliqua exusta in cineres convertuntur. Alii Daemonem esse, qui figuram demortui hominis induerit, opinantur, sub eaque homines, quos ipse vult, conficere. Hanc opinationem ex populi animis evellere conati sunt, non veteres modo, (neque enim nupera est, aut hodie nata in Græcia, haec opinio,) sed etiam recentiores pii homines, qui Christianis a confessionibus sunt."

At p. 149. "Graeci, cum similia vident corpora, quae post obitum in coemeteriis indissoluta comperiuntur, et tympani more extensa cute tumescunt, excommunicatorum esse corpora fatentur, eaque post absolutionem statim dissolvi."

Again, p. 151. "Et stultissima ista de excommunicatorum cadaveribus indissolutis opinatio apud eam nationem adeo invaluit, ut jam nemo sit, si uspiam tale quid comperiatur, qui dubitet, cadaver illud esse excommunicati; quod extrahunt, et variis deprecationibus ac dicendi formulis absolvunt; ordinemque, in tali absolutione servandum, idem Nomocanon, cap. LXXXII. exactissime prosequitur; quibus exsolutis, asseveranter dicunt, cadaver subito in cineres converti. Ipse nihil tale uspiam in Graecia vidi. Audivi tamen saepius ab Athanasio, Imbri Metropolita, homine frugi, et qui rem factam mendacio non contaminasset, cum Ihasi[1] moraretur in regione Theologi extra urbem in ecclesia sancti Georgii, exoratum a civibus, ut super cadavera, quae ibidem pleraque conspiciebantur indissoluta, excommunicationis absolutionem recitaret, morem gessisse, neque dum finita absolutione cadavera ea omnia in pulverem abiisse. Referebat idem, Constantini cujusdam cognomento Rezepii, qui ex Turca Christo nomina dederat, quod vitam viveret omnibus flagitiis inquinatam, excommunicati, corpus tumulatum fuisse in ecclesia sanctorum Petri et Pauli nationis Graecorum Neapoli, et per plures annos indissolutum mansisse; ab eodem postea, et aliis Metropolitis duobus, Athanasio Cypri, et Chrysantho Lacedaemonio, benedictione subactum, ut alia mortuorum cadavera, pulverem factum. Et quod admirabilius est, dum Raphaël in Patriarchatu praesideret, eodem asserente, qui alium excommunicaverat, postea, compulsu Daemonis, Christum ejurat, excommunicati licet Christiani demortui cadaver mansit indissolutum. De eo certior factus Patriarcha, accersitum Turcam, qui excommunicaverat, monet, ut absolutionem impertiatur. Primum ille renuere, factum detestari; nihil esse Turcis cum Christiana religione commune, dicere; quare Christiani Christianum absolverent: cum vero pertinacius exoraretur, obedit, et absolutionem super excommunicatum recitat; oculatum se testem fatetur, qui perhibet; prope finem absolutionis, cadaveris tumor cessat; et in cineres omnia convertuntur. Rei novitate Turca attonitus, ad magistratum ocius advolat, rem uti facta est narrat, edicit omnibus veram religionem Christianam, quam ipse per summum nefas deseruerat; eam se denuo amplecti, Mahumetanam detestari. Monetur a Turcis, ut sapiat, ne tormentis se objiciat. Ille, se Christianum velle mori, contendit. Quid plura? pertinacem condemnant: ducitur ad supplicium, et Christianam religionem praedicans, morti, summo supplicio mactatus, deditur.

"Plura de his excommunicatis narrat Christophorus Angelus, de vita et moribus recentiorum Graecorum cap. XXV. Non praeteribo vero, quod ille ex Cassiano Historico recitat. Ὁ δὲ νοῦς οὕτως ἔχει. Οὗτος ὁ Κασσιανὸς ἐστὶν Ἕλλην παλαιὸς ἱστορικός, καὶ γράφει ἐν ταῖς ἱστορίαις αὐτοῦ, ὅτι ἐγένετο ποτὲ μερικὴ σύνοδος ἔν τινι τόπῳ ἑκατὸν ἐπισκόπων, καὶ πάντες ὀρθῶς ἔλεγον. εἷς δὲ ἐξ αὐτῶν ἐναντιοῦτο πᾶσι. Τότε ἐκεῖνοι ἀναθεμάτισαν ἐκεῖνον, καὶ οὕτω τέθνηκεν ἀφωρισμένος. καὶ ἔμεινε τὸ σῶμα ἐκείνου ἑκατὸν ἔτη δεδεμένον, ὡς σίδηρον. μετὰ δὲ ἑκατὸν ἔτη ἐγένετο πάλιν σύνοδος μερικὴ ἐκεῖσε ἑκατὸν ἐπισκόπων. Τότε ἐκεῖνοι οἱ ἐπίσκοποι εἶπον ἀλλήλοις, ὁ ἀναθεματισμένος ἐπίσκοπος ἐκεῖνος ἥμαρτεν εἰς τὴν ἐκκλησίαν, καὶ ἀφώρισεν ἐκεῖνον ἡ ἐκκλησία. Τὸ λοιπὸν καὶ ἡμεῖς ἐκκλησία ἐσμέν, καὶ συγχωρήσωμεν ἐκεῖνον, ἐπειδὴ ἀνθρώπινόν ἐστι τὸ ἁμαρτάνειν· καὶ

[1] Qu. Thasi?

συνεχώρησαν ἐκεῖνον, καὶ αὐτίκα μετὰ τὴν προσευχὴν ἐστράφη εἰς κόνιν ὁ ἑκατὸν ἐτῶν ἄλυτος."

The story told in RICAUT, State of the Greek Church, (in 1678) p. 278. foll. is highly illustrative of the superstition. It is manifest that the operations mentioned above in note 22, p. 205. as performed on the dead body by the Morlacchians, are simply to prevent its becoming a vampire, and are not at all designed as an indignity. Still less was any indignity meant in such cases as the following. "Cum aliquis femina parere debet, et non potest, in ipso dolore si mortem obierit, in ipso sepulchro matrem cum infante *palo in terram transfigunt.*" BURCHARD OF WORMS, Samlung der Decrete, Colon. 1548. quoted by J. GRIMM, Deutsche Mythologie, p. XL. Doubtless this was done to prevent their becoming vampires. Again, in the same collection, p. 200. b. (in GRIMM, p. XXXIX.) "Fecisti quod quaedam mulieres instinctu diaboli facere solent, cum aliquis infans *sine baptismo* mortuus fuerit, tollunt cadaver parvuli, et ponunt in aliquo secreto loco, et *palo corpusculum transfigunt, dicentes, si sic non fecissent, quod infantulus surgeret et multos laedere posset.*" The stake therefore, in all probability, was driven through the bodies of such persons as could not be buried in consecrated ground, and were thus thought likely to become vampires.

Now suicide, though not expressly prohibited in Scripture, and though frequently practised by Christians of the early centuries[2], was considered by the Church, from the time of Augustine downwards, as a great crime; and different Councils decided, that no masses should be said for the souls of such as laid violent hands on themselves, and that the usual funeral service should not be read over their remains, which were no longer buried in consecrated ground.

I think this investigation affords a satisfactory explanation of the origin of the immemorial English custom of driving a stake through the bodies

[2] Not only by the Donatists, who considered suicide as a martyrdom, and made it, to use Augustine's words, their "daily sport," "ludus quotidianus;" but by a great many females, who thus aimed at acquiring the crown of virginity, "on which so high a value was set in the early ages of Christianity." "Eusebius and other ecclesiastical writers mention many Christian women who put themselves to death, by drowning, leaping from precipices, or other ways, when the confusion and trouble of the times threatened them with violation: concerning whose voluntary suicide some of the fathers speak with great tenderness, if not with a degree of approbation; and many of these virgin-suicides were admitted into the calendar of saints. 'She aimed,' (says Chrysostom of Pelagia,) 'both at the crown of martyrdom and the crown of virginity; but there being so great a hazard of losing the latter, she had *just cause* to prevent so great an injury to herself by a previous voluntary death.' Ambrose also bestows the highest encomiums on this action of Pelagia, and zealously commends the behaviour of those women who killed themselves to avoid the hazard of violation. Jerome likewise excepts the case of preserving chastity from his general censure of suicide." MOORE's Full Inquiry, Vol. I. pp. 293—295.

of such persons, which usage, however, never formed any part of the prescribed legal punishment of suicide, and was not mentioned in the coroner's warrant for an ignominious burial. In all likelihood it arose from a belief, similar to that with which the traveller in Dalmatia, and the collection of Burchard, make us acquainted; and was therefore not at all meant as an indignity, but simply as a precaution, which, in consequence of the general belief in vampires prevalent in England, must have been thought likely to contribute to the mutual benefit of both the dead man and his survivors, by ensuring, to the former, the quiet occupation of his tomb, and, to the latter, freedom from molestation by his nightly rambles.

The Commentator on the Laws of England (BLACKSTONE, Vol. IV. p. 189) seems to consider that our law "wisely and religiously" acted on both the reputation and fortune of the *felo de se:* "on the former, by an ignominious burial in the highway with a stake driven through his body; on the latter, by a forfeiture of all his goods and chattels to the king." The feelings of the nineteenth century have however made a change in what the law had so wisely contrived. By an act of Parliament, 4 Geo. IV. cap. 52. intitled "An act to alter and amend the law relating to the interment of the remains of any person found *Felo de se*," it is enacted, that the coroner or other officer "shall give directions for the private interment of the remains of such person *Felo de se*, without any stake being driven through the body of such person, in the churchyard," &c. It is nevertheless expressly provided, that nothing in the act "shall authorize the performing of any of the rites of Christian burial on the interment of the remains of any such person;" and the burial is only to take place between nine and twelve o'clock at night.

There is only one point left unexplained: why the place commonly chosen for the English suicide's burial was not simply by the way-side, but where four roads meet. I conceive that, in the superstitious times when the rite originated, such places were ordinarily regarded as *the most holy* that existed out of consecrated ground; and therefore the humanity of our ancestors, towards the unfortunate suicide, was as much shewn by the locality which they selected for his burial-place, as by the rites which they observed in interring him. It was especially at cross-roads that sepulchral monuments were erected, by the piety or superstition of our Roman Catholic forefathers, who thus secured the prayers of passers by in favour of the dead [3]. Moreover, "In early times crosses were erected at most places of public

[3] See BORLASE, Antiquities of Cornwall, B. IV. ch. XII. p. 362. where several examples of these sepulchral monuments are given, Compare also the ARCHAEOLOGIA, Vol. XIII. p. 216. They continued to be erected, in Scotland and Ireland, till the beginning of the last century.

concourse, or at the meeting of three or four roads or high-ways[4]:" and we know that every such place was thus rendered so holy as even to afford an asylum for criminals, equally with a church itself[5]. Thus a cross was probably generally standing at the place so kindly chosen for the suicide's interment, and would serve "to put devout people in mind to pray for the soul" of him who was buried there. At the present day, the Roman Catholic peasantry of the sister kingdom, when they go in procession with a dead body, still manifest a religious respect for the sanctity of cross-roads. See CROKER, Researches in the South of Ireland, p. 172. "At every cross-road, there is a general halt; the men uncover their heads, and a prayer is offered up for the soul of their departed chief." It would lead us too far from the subject of the Cretan superstition, if we were to endeavour to trace the origin of this modern Christian belief, which would perhaps be found to spring from a pagan source. Diana Trivia may, however, be mentioned; and, on the mystical Pythagorean letter, and the three-ways, of both the Heathen and the Christian Orcus, the reader may consult LOBECK, Aglaoph. pp. 1342—1344.

[4] ASTLE, on Stone Pillars, in the Archaeologia, Vol. XIII. p. 216.

[5] "Si quis ad aliquem crucem in via, persequentibus inimicis, confugerit, liber ac si in ipsa ecclesia, permaneat." DUCANGE, Gloss. med. et. inf. Lat. Tom. II. 1184. ARCHAEOLOGIA, Vol. VIII. p. 28.

NOTE C.

The original Greek of the story told at pp. 217–218. is as follows.

Μία φόλα[1] *μου ἔκαμε νάκλι*[2] *ἕνας ἄνθρωπος ὅπως ἤσασι παγομένοι*[3] *δύω ἄνθρωποι στὰ 'ψηλότερα βουνά, ἀποῦ κατοίκουσιν ἀγρίμια*[4]*, καὶ ἐκαθίσασι τὴν νύκτα μὲ τὸ φεγγάρι εἰς ἕνα διασέλι διὰ νὰ κυνηγήσουσι τὰ ἀγρίμια. καὶ ἐκεῖ ἀκούσασι τραβαίλιον*[5] *πολύ, καὶ ἐθαῤῥέψασιν ὅπως νὰ ἦσαν ἄνθρωποι*

[1] Φόλα i. e. *φορά*.

[2] Νάκλι, from the Turkish نقل *nákl*, recital, narration, *διήγησις*.

[3] Παγομένοι, from *ὑπάγομαι*.

[4] Ἀγρίμια. The wild-goats are thus called. The word is used by an anonymous author, quoted by DUCANGE, Glossar. Graec. 19.

Ἀγρίμια, λάφια, καὶ λαγούς, πλατόνια μουζομῆτες
ἐπίαναν μετὰ δίκτυα, καὶ μὴ σκυλία οἱ τοπίτες.

The word *ἀγριμαῖον*, in ancient Greek, was used to denote all kinds of game, or wild animals the object of the chase, as opposed to tame ones. KORAY, ATAKTA, Vol. II. p. 12. Ἀγρίμι, *ζῶον ἄγριον. ἀπὸ τὸ Ἀγρίμιον, καὶ τοῦτο πάλιν ἀπὸ τὸ* Ἑλληνικ. Ἀγριμαῖον, *κατὰ τροπὴν τῆς διφθόγγου* αι *εἰς* ι, *τροπὴν συνήθη καὶ εἰς ἄλλα πολλά.* Εἰς *τὸν* Ἀθήναιον (XII. p. 549. f.) Ἀγριμαῖα *λέγονται, τὰ ἀπὸ κυνηγέσιον ἀγρευμένα ἄγρια ζῶα* (venaison) *χρήσιμα εἰς τροφήν.* Compare Vol. III. under the words *Ἀγρίμιον* and Πλατώνιον.

[5] Τραβαίλιον, the Italian *travaglio*.

διὰ νὰ φορτώσωσι χιόνι, διὰ νὰ τὸ πάγουσι στὰ Χανιά. Καὶ ἀπ' ἧς[6] *ἐσιμώσασι*[7] *κοντήτερα γροικοῦσι*[8] *βιόλιαις, καὶ λύραις καὶ ῥογήσιμα*[9] *παιγνίδια, καὶ δὲν εἶχαν ποτὲ ἠκουσμένα τέτοια παιγνίδια· διὰ ταῦτος ἐγνώρισαν καὶ αὐτοί, ὅπως δὲν ἦσανι ἄνθρωποι, ἀῤῥὰ ἦσαν δαιμονικὸ συνέδριον· καὶ τοὺς θώρουσι, καὶ πέρνουσι ἀπὸ κοντὸ διάστημα ἀπὸ ἐκεῖ ἀποῦ ἐκάθοντο, μὲ ῥογιορογῆς·*[10] *φορέματα, καὶ ἄῤῥοι μὲ ψάρα ἄρογα, ἄῤῥοι μὲ ῥογήσιμα, καὶ τοὺς ἔδειξεν ὅπως ἦσασι καὶ γυναῖκες καὶ ἄνδρες, πέζοι καὶ καβαλλάροι*[11]*, πλῆθος πρᾶμμα. καὶ οἱ ἄνδρες ἤσασιν ἄσπροι*[12] *ὡς τὰ περιστέρια, καὶ αἱ γυναῖκες*

[6] Ἀπ' ἧς. Koray does not seem to have been aware of the existence of this phrase instead of the common *ἀπῆν*. If his conjecture respecting the origin of the expression *ἀπῆν* is correct, and it seems to be so, we may consider this *ἀπ' ἧς* as the more ancient form. I write it as I heard it *ἀπ' ἧς* and not *ἀφ' ἧς*. The aspirate is never pronounced by the modern Greeks, (see Vol. I. p. 97.) and therefore the *π* naturally retains its place. Koray's words are: *τὸ Ἀπῆν (ἀντὶ τοῦ Ἀφοῦ,) ἐνδέχεται νὰ ἐφθάρῃ ἀπὸ τό, Ἀφ' ἧς, ἐλλειπτ· ἀντὶ τοῦ Ἀφ' ἧς ὥρας ἢ ἡμέρας (ὡς καὶ τὸ Ἀφ' οὗ, εἶναι Ἀφ' οὗ χρόνου,) τὸ ὁποῖον ἐμετεβάλθη ἔπειτα εἰς αἰτιατικήν, Ἀφ' ἥν, Ἀπήν, διὰ τὴν ἐπικρατήσασαν συνήθη χρῆσιν τῆς πτώσεως ταύτης ἀντὶ τῆς γενικῆς.*

[7] Ἐσιμώσασι. The opposite to *σιμόνω* is *ἀποσιμόω*: THUCYDIDES, IV. 25. says, *ἀποσιμωσάντων ἐκείνων*, in this opposite sense. See the observations made by KORAY, ATAKTA, Vol. IV. pp. 498-9.

[8] Γροικῶ or Δροικῶ, for both forms are used, is a very common word in Crete, where its synonym *ἀκούω* is hardly ever heard. *δὲν δροικᾷς*, for instance, is always used instead of *δὲν ἀκούεις*. In other parts I believe the commoner form is *γροικῶ*. The etymology of the word is satisfactorily explained by KORAY: *Ἡ χυδαιότης, ἀκούουσα τὸ Ἀγροικίζομαι, ἐνόμισε τὸ ᾱ στερητικόν, καὶ ἀκολούθως ἐσυμπέρανε ὅτι τὸ χωρὶς τῆς στερήσεως Γροικίζω, Γροικίζομαι, Γροικῶ, ἔπρεπε νὰ σημαίνῃ, ἐναντίον τι τοῦ Ἀγροικίζομαι, ἤγουν νοῶ, καταλαμβάνω.* ATAKTA, Vol. II. p. 95.

[9] Ῥογήσιμα, and soon after *ἀῤῥά, ἀῤῥοί, ποῤῥά*, are all obtained by the substitution of *ρ* in the place of *λ*.

[10] Ῥογιορογῆς. In common Greek we have *μίας λογῆς*, *of one sort*, *πολλῶν λογιῶν*, *of many sorts*, and so on. The Sfakian word *ῥογιορογῆς* seems to be for *λογιολογῆς*, and to have nearly the same meaning as *πολλῶν λογιῶν*.

[11] Καβαλλάρης, a rider. On the usage of *καβάλλης* by the old Greek writers, see VOSSIUS and ALBERTI on Hesychius, who explains the word *καβάλλης* as meaning *ἐργάτης ἵππος*. KORAY, (ATAKTA, Vol. IV. p. 188. v. ΚΑΒΑ'ΛΑ,) quotes from Plutarch, "*Ὄνῳ τινὶ τῷ τυχόντι καὶ καβάλλῃ χρώμενος.*" The word became the *Caballus* of the Romans, (which is still *caballo* in many parts of Italy,) on which see KOENIG, on Persius, Prol. v. 1. I think I remember to have heard *καβάλλο* used in some island of the Archipelago: at all events it is not found in Crete, where the common *ἄλογον* and the peculiar *κτῆμα*, on which see Vol. I. p. 82. are alone used.

[12] Ἄσπροι. On the etymology of the word, see Vol. I. p. 52. This *ἀσπρότης*, or fairness of complexion, is esteemed as the greatest element of personal beauty by the Greeks of both sexes. Captain Maniás himself used to amuse me with his complaints of becoming *μαυρισμένος* by the sun, during my long rides throughout the day: and [an Englishman who travels in Crete may often overhear such exclamations of wonder and admition as *τί ἄσπρο πρόσωπον*! from some of the fair villagers, who every where flock around

πορρὰ ὡραίαις, ὡς αἱ ἀκτῖνες τοῦ ἡλίου· εἴδασι καὶ ἐβαστοῦσαν ἕνα πρᾶμμα ὡς καθῶς καὶ βαστοῦνι τὸν ἀπεθαμένον νεκρόν· μεταῦτος ἀποφάσισαν καὶ αὐτοὶ νὰ τοὺς βαρέσσουσι καμμίαν μπαρουτιά, ἀπ' ἧς ἐμπρὸς περάσανι· εἶχαν καὶ ἕνα τραγοῦδι ἀκουσμένο, καὶ ἐλέγασιν ὅπως

Νύμφην πά νὰ πάρωμε, νύφην κύραν,
ἀπὸ τὸ κρημνισμένο, νύφην μοναχήν·

καὶ ἀποφασίσασι καὶ τζὴ ἐπαίξασι μία μπαρουθιά· διὰ ταῦτος καὶ αὐτοὶ ἐφωνάξασι, οἱ προσθινοί, μίᾳ φωνῇ, "τί εἶναι;" καὶ οἱ ὀπισινοὶ ἀποκριθήκανι, "Τὸν γαμπρόν μας ἐσκοτώσασι, τὸν γαμπρόν μας ἐσκοτώσασι" · καὶ ἔκλαιγαν, καὶ ἐφωνιάξασι καὶ ἐφύγασιν.

around him, on his dismounting for the evening, to learn what can have induced him to visit the valleys and mountains of Crete.

ANCIENT WALLS NEAR ANOPOLIS.

CHAPTER XXXVII.

THE APPEARANCE, IN SFAKIA, OF A SAINT AND PROPHET, WHO WORKED MIRACLES. THE MARRIAGE OF COUSINS NOT ALLOWED IN THE GREEK CHURCH. THE SAINT'S REPUTATION AND SUCCESS CONFINED TO THE MOUNTAINEERS. VESTIGES OF AN ANCIENT CITY: MASSIVE REMAINS OF POLYGONAL MASONRY. PRODUCE OF THE PLAIN AND MOUNTAINS OF ANOPOLIS. SELF-DESTRUCTION OF A CHRISTIAN MOTHER, WITH HER CHILD. A SFAKIAN WORTHY. SFAKIAN CUSTOMS. PUNISHMENT OF A WIFE IF SUPPOSED FAITHLESS. FEUDS PERPETUATED BETWEEN FAMILIES.

In a country where the belief in such supernatural occurrences, as have been described in the previous chapter, is not confined to old women and children, but prevails among the whole population, it is not surprising to learn how powerful an influence has occasionally been exerted over the people by artful men, who have played on their simplicity and credulity, and have made the profession of a Saint and Prophet answer well their own base ends. The older Sfakians remember that not long after the devastation of their country in 1769, a saint, who was stone-blind and never tasted any food[1], came from one of the monasteries of the Holy Mountain[2] to preach and prophesy to the Sfakians. When surrounded by a numerous audience he would sometimes declare who were the sinners, and who the righteous men among his hearers. Once some Mohammedans joined his congregation, and scarcely had they entered the church before he exclaimed, that they would never learn any thing from him but must go to their prophet Mohammed. He declared that the wrath of God could only be averted from the land by their abandoning their unchristian and murderous practices: and the influence of his reputation and preaching among them was confessedly great enough to put a stop, for awhile, to the murders which from time immemorial were so common in Sfakiá. But the date of this worthy man's ministry is so remote, that we do not learn so much about him as we may of another saint, who is still living[3], and who made his appearance in Sfakiá

[1] My Sfakian informant, although he had been a leader in the whole war, and had lived in exile at Nauplia ever since his country was given up to Mehmét-Alí, was firmly persuaded that the *ἅγιος γέρωντας* never eat anything. Many examples "of wonderful fasting" are stated in JONSTONUS, An History of the Wonderful Things of Nature, Classis x. Ch. 2. §. 7. p. 315. ed. Lond. 1657. The periods vary from the modest length of forty days to that of forty years! "Hermolaus knew a priest who lived in health forty years, without any thing, but by sucking in the air." The goats of Cephallenia are said always to have satisfied their *thirst* in the same way: see the Book de Mirab. ausc. n. 9. p. 831. ed. Bekk. *Αἱ ἐν Κεφαλληνίᾳ αἶγες οὐ πίνουσιν, ὡς ἔοικεν, ὥσπερ καὶ τἄλλα τετράποδα, καθ' ἡμέραν δὲ πρὸς τὸ πέλαγος ἀντία τά πρόσωπα ποιήσασαι χάσκουσιν εἰσδεχόμεναι τὰ πνεύματα.*

[2] Τὸ *Ἅγιον Ὄρος*, as Mount Athos is called by the Greeks.

[3] I saw him at Nauplia in 1834. He has long since abandoned his sacred profession spoken of in the text.

during the year 1811. His dress consisted of a single coarse garment, under which he wore an iron chain hung about his neck. He preached repentance and newness of life[4], exposing to the wild mountaineers, by whom he was surrounded, the wickedness of their ways, dwelling chiefly on the daily acts of violence, robbery and murder, which they were committing, and declaring to them, that the anger of Heaven was awakened, and would soon shew itself in earthquakes, pestilence and war, to the desolation of the whole land, unless they repented and changed their course of life. Every one went to hear him: his addresses were always vehement, and lasted several hours: he was regarded as a man sent by God, (he professed a divine mission,) and the impression he produced was very great. His fare was most simple, and abstemious almost to starvation: he ate no meat, eggs, fish, cheese or milk: a few herbs, simply boiled, and a crust of black barley-bread, formed his single daily repast[5]. He professed not to receive any money from his hearers: nevertheless, at the end of his discourse, he used to tell the people, that although his reward was not of this world, still he had left a poor monastery on the Holy Mountain, where six hundred

[4] The object of Venturius of Bergamo, in his preaching throughout Lombardy and Tuscany in 1334, was "*to bring sinners to repentance*; and so great was the success, and so visible were the fruits of his eloquence, that more than 10,000 Lombards, of whom many were of the higher ranks, set out to pass the season of Lent at Rome." WADDINGTON, History of the Church, Vol. III. p. 112. Note at the end of Ch. XXIII. Nearer home we had, about a hundred and sixty years ago, one John Exham, an early religious associate of William Penn, who was distinguished by the name of the Quaker-Prophet. "His enthusiasm was so great that he walked through the streets, his head covered with sackcloth and ashes, preaching repentance and amendment of life,—for which he suffered a long and severe imprisonment." CROKER, Researches in the South of Ireland, Ch. X. p. 198.

[5] It is said of one Athanasius, a contemporary of Nicephorus Phocas, who used to wear an iron collar round his neck, to which a wooden cross weighing six or eight pounds was fastened, that "he seldom ate above three or four times a week, and once lived seven days without meat or drink; and once nine days, excepting only that he ate and drank the Holy Sacrament twice in that time." RICAUT, Greek Church, p. 234.

kalógheri[6] were praying and fasting, daily and hourly, for the sins of a wicked world, interceding with the Deity to obtain a remission of the punishment which mankind merited, and therefore those who had many sheep might give one, or even two, for repairing this monastery. Now, almost every one in Sfakiá had *some* sheep, if not a whole flock, and thus the Prophet-saint obtained, perhaps, a much greater sum than, without such a suggestion, he would ever have raised by any voluntary contribution. The devastation committed among the Sfakian flocks, by this ravenous monster, as he prowled through the mountains, reminds me of the words of our poet[7]:

> Wolves shall succeed to teachers, grievous wolves,
> Who all the sacred mysteries of Heaven
> To their own vile advantages shall turn
> Of lucre and ambition[8].

So holy a heaven-sent messenger, as the saint was believed to be, did not remain long among this simple-minded people without their seeing *proofs* of his divine mission. The shepherd's wife, in a house where he lived, was once preparing some eggs for her family's meal: after breaking three, she went to the door for an instant, and, on her return to the fire, found four in the pan! "But how is this, holy priest? I only broke three." "I have blessed them," he replied[9],

[6] There is a very good account of the monasteries of Mount Athos in Col. LEAKE's Travels in Northern Greece, Vol. III. pp. 114—142.

[7] MILTON, Paradise Lost, XII. 508.

[8] With the Cretan saint's pecuniary success we may compare that of a more recent impostor, who, even amid the din of arms of the revolution, exhorted his auditors, among other things, "to use no other weapon than the sign of the cross." A reverend English missionary says of him, "I should myself have been rather disposed to consider the man as deceived, than a deceiver, were it not for the immense sums of money which he amassed, by means of his preaching. I have been assured that he collected no less than 700,000 piastres, the offerings of enthusiastic multitudes." HARTLEY, Researches in Greece, p. 357.

[9] Τὰ εὐλόγησα.

"and they are become four." Her husband soon returned, and remained incredulous, refusing all assent to his wife's opinion of the holy hermit's miraculous powers. Their guest left them before day-break, and, soon after his departure, his host missed a counterpane. "Ah, so this is your holy man! he has stolen our páploma." The Sfakian pursued their unfortunate guest, whom he thought a thief, while his wife and all the women regarded him as a saint. When within hearing, though still at a considerable distance from the object of his pursuit, the Sfakian called out, demanding restitution of the páploma. The holy man immediately stopped, detached the counterpane from his mule, and, when his late host overtook him, gave it up, declaring, however, that assuredly "God would not permit him ever to use it again, but would send fire from heaven and consume it." The hard-hearted mountaineer heeded not the threat, but took the counterpane, and had not proceeded half a mile with it on his shoulder, before there suddenly burst out from it a blaze of fire[10].

After traversing all the Sfakian mountains, the thavmatúrgos proceeded to preach on their lower slopes, and when he held forth at Thériso, the church was crowded even soon after midnight. At an hour and a half before daybreak[11], he entered, and, after waiting awhile, he asked his audience if they were all assembled: they exclaimed, all: even the shepherds had left their flocks unattended on the mountains: all were come, from far and near, to hear him. He now commenced his address[12], which had continued for near an hour, (its whole length being generally three to four hours,) when an unlucky wight entered the church. The holy man,

[10] Many such saints are now flourishing in Afghanistan, "and the ignorance of their countrymen ascribes to them the gift of prophetic dreams and visions, and the power of working miracles." See Mr ELPHINSTONE'S admirable work on the kingdom of Caubul, p. 220.

[11] See above, p. 164.

[12] Διδαχή.

immediately on seeing him, exclaimed, "Woe unto thee! art thou come to the house of God at this hour? I liken thee to the traitor Judas." Now this youth had committed the crime of having married his second cousin, and the authority of the church had been exercised to effect their disunion[13]; but they found themselves unable to live apart:

> Once again
> The persevering wedge of tyranny
> Achieved their separation;—and once more
> Were they united—to be yet again
> Disparted—pitiable lot!

They were ready to bear even that "immortality of ill," which, as they must have supposed, would follow their transgression of the ordinary custom of their country, and of the commands of those who spoke in the name and with the authority of religion. Strong, indeed, must have been the affections of the ill-starred couple, to enable them thus to turn a deaf ear to the denunciations of their priests:

> Quanti dolci pensier, quanto desio,
> Menò costoro al doloroso passo!

The unfortunate man in question was considered as a godless reprobate by all the villagers of his neighbourhood, and thus the pretended prophet's reputation for sanctity and omniscience, was greatly increased by the circumstance of this fortuitous address.

One remarkable consequence of the holy man's preaching in Sfakiá, was a degree of order which none of the inhabitants had ever known: scarcely ever was a murder heard of, and robberies became very rare: I am

[13] HARTLEY, Researches in Greece and the Levant, p. 79. "A story was recently current in the Archipelago, of two cousins who had formed a most sincere attachment for each other. By some means, probably by a considerable bribe, they had prevailed on a priest to solemnize matrimony between them. They had lived happy and united for two or three years; when, at length, the marriage was declared null and void by superior authority, and they were forcibly separated."

told that he thus saved the lives of at least a hundred souls during the next five or six years. This order[14] lasted till nearly the commencement of the revolution, when the bonds of Christian brotherhood and union against a common foe prevented the revival of old feuds: and thus, since 1812, such exactions of vengeance[15] as wiped out the offence in the blood of the offender, and as used previously to be of daily occurrence, have scarcely ever happened[16].

Our saint, after converting from their evil ways the Sfakians and Rhizítes, went into the city of Khaniá: but there his eloquence was powerless, and the inhabitants believed not. When the Sfakians and Rhizítes learnt this, they exclaimed, "Ah, so it was when Christ himself preached: the plain and simple peasants believed, but the great and the learned[17], the Pharisees, believed not."

May 6.

We must now leave the modern inhabitants of these mountains, to examine the remains left of a city built by their ancestors. An ascent of a few minutes up the rocky elevation at the foot of which the hamlet of Rhíza is built, brings us to the site of the ancient city, from which we see along the southern coast as far as the projecting point of Mesará. Fránko-kástello, the scene of Khadjí Mikháli's fatal contest, and of his death, is about twelve miles off: the white-washed church of the Panaghía, just by the castle, is a very distinct object:

[14] Εὐταξία. [15] Ἐκδικήσεις.

[16] Similar effects were produced in Italy, in 1399, by the preaching of the White Penitents. "Every one pardoned his neighbour, and dismissed the recollection of past offences. Enmities which no ordinary means could have reconciled were put asleep. It was a festivity of general reconciliation. Ambuscades, assassinations, and all other crimes were for the season suspended." See the note already referred to, and contained in the 3d vol. of the History of the Church, by my learned friend Mr WADDINGTON; to whom I am indebted for the indication of this interesting passage of his work.

[17] Οἱ μεγάλοι καὶ γραμματισμένοι.

below us is the port Lutró, and its little village, the winter residence of nearly all the Anopolítans: it seems to be about a mile off: I am, however, assured that following the road which leads to it, its distance is full three miles.

The whole circumference of this rocky elevation, occupied by the ancient city, is about a mile; perhaps hardly so much. To the south I observed no traces of walls: the rocks being nearly perpendicular on that side, it is very unlikely that walls should ever have been built there. The chief remains are on the west; where a considerable piece of ancient walling, of the very earliest style, still exists. Its length is about three hundred paces, and its width is generally about six feet. Its height varies from five to eleven feet. The chisel has been nowhere used on any of the stones. The sketch at the head of this chapter is, like those which have preceded it, a faithful representation of the part which it professes to exhibit[18]: if there is any difference between the rest of the walls and this specimen, it consists in the fact, that the proportion of small stones among the very large ones is somewhat greater in the wall generally, than it is in this portion of it.

The plain, on which are situated the hamlets comprehended under the common name of Anópolis, is covered with patches only of cultivation: its length is about a mile and a half, its breadth perhaps three quarters of a mile: to the south it is bounded by this rock, and to the north by the highest range of the Sfakian mountains.

One sees, on looking down on the hamlets of Anópolis, how much greater desolation has been caused here than at Askýfo by the war. The great number of these burnt and ruined houses, is partly owing to the fact, that a considerable portion, I am assured more than fifty families, of the surviving Anopolítans, expatriated themselves, on finding that, in spite of all the exertions

[18] And which is eleven feet nine inches in length, and nine feet six inches high.

of the Cretan Greeks, their country was still to remain subject to the Turks, and they are now scattered through the islands of the Archipelago. I myself saw several of them, at Mélos, in the autumn of 1833. I am assured that, before the revolution, there were not less than five hundred and eighty armed men of this village.

On this ancient site stones are gathered together in heaps, sufficiently great and numerous to serve for building a city: it would seem, that, before the revolution, the Sfakians suffered not an inch of arable land to remain uncultivated. Barley is still the main produce of the plain of Anópolis, in which there also were, before the arrival of Khuseín-bey's troops in 1824, upwards of two thousand olive-trees: he remained here seventeen days, with many thousand troops, in the month of March, and thus the greater part of the olive-trees were converted into firewood. Of the fig-trees also, which existed at that time, some have sprung up again, but the olives were all cut so low that none of them recovered.

The flocks of sheep and goats which browze on the mountains from April till October, and are taken down to the shore near Lutró, or elsewhere, for the rest of the year, supply the Anopolítans with milk, cheese and meat: the cultivated patches of their plain produce some corn and wine, though not enough for their consumption. They have also some mulberry-trees, and thus obtain a little silk: their oil is wholly imported: their principal export is cheese.

After examining the ancient remains, among which there are said to be many cisterns, the entrances to which are, however, blocked up, I return to my host's cottage, and hear some more stories from his talkative wife.

I learnt, elsewhere, an event which happened here during the war. When the Turkish force was at Murí, on its way to Anópolis, they captured, along with other persons, a young mother and her infant, whom she carried in her arms. She was beautiful enough to be an object of contention among those who laid claim to the spoil; and, while her brutal captors, when at Anópolis, were quarrelling as to who should possess her, she went out, with her child in her arms, to one of the large open wells from which the villagers draw their water, and, plunging into it, escaped all the horrors of slavery by a voluntary death[19]. Such instances as this, give to the war, which was waged for so many years between the Christians and Mohammedans, its true character: as to the men, it was one of extermination; while, as to the women, it was even worse: and this poor Sfakian mother, like many other Christian women in different parts of Crete and elsewhere, shewed that she thought death a happier lot than that which was reserved for her by the enemies of her religion[20].

At half past twelve I set off, and, crossing the plain in a west-north-westerly direction, I soon arrive at the hamlet of Hághios Dhemétrios, where I found one of the Anopolítan exiles, who played an important part in the latter events of the war in Crete, and has returned from Nauplia, where he at present dwells, in order to

[19] Thus also twenty-two Suliote women threw themselves from the rocks, near Kamarína, in preference to falling into the hands of the enemy, and "several of the women, who had infants, were seen to throw them over, before they took the fatal leap:" LEAKE, Travels in Northern Greece, Vol. I. p. 246. Compare FAURIEL, Tom. I. p. 302.

Σκλάβαις Τουρκῶν μὴ ζήσωμεν·
παιδιά, μαζή μ' ἐλᾶτε!

Nevertheless, such is the tenacity with which we cling to life, that even slavery has generally been preferred to death: M. d'Herculez, Austrian Consul at Khaniá, writing to his government, on the 5th of October, says, "Chaque jour on emmène au marché des malheureuses femmes Grècques, que la faim chasse des montagnes, et qui prefèrent l'esclavage à la mort."

[20] A Greek woman exclaims, in a well-known song, FAURIEL, Tom. I. p. 138.

Κάλλια νὰ ἰδῶ τὸ αἷμα μου
τὴν γῆν νὰ κοκκινίσῃ,
παρὰ νὰ ἰδῶ τὰ μάτια μου
Τοῦρκος νὰ τὰ φιλήσῃ.

Better that I should see my blood
With crimson stain the ground,
Than that I e'er should see my eyes
Kissed by a Turkish hound.

dispose of his property in Sfakiá, being unable to persuade himself to live any longer in the land of his fathers, "now that it is to remain under the Turkish yoke." His name is Manúsos Vardhulákes: he is of a powerful muscular frame, and seems to be as hard as iron: in all probability he is near sixty years of age. Although I resisted his importunity, and escaped staying the greater part of the day with him, as he wished me to do, yet, having on a subsequent occasion seen a good deal of him in the place of his exile, Nauplia, I will detain my reader for a moment with the detail of a few passages in my friend's history, which serve to illustrate the modes of thinking and acting in this district some years ago.

The ancient worthy, who is certainly a good specimen of the wild Sfakian mountaineer of former times, was one of the leaders of the Greek forces when they ravaged Setía in 1829. He admits that he has killed or wounded about ten men in private quarrels in the old times. "Only few people," he adds, "ever died a natural death in Sfakiá;" and when a man was slain it became the duty of his family to avenge his death: if forty or fifty years elapsed before an opportunity offered itself, the obligation still lost none of its force on account of the delay. The debt remained due, and it was a point of honour to exact its payment. Old Manúsos tells me simply that such was the custom[21] of

[21] Ἤτουι τὸ συνήθειον μᾶς. Two hundred and sixty years ago, according to Foscarini, the Sfakians used never to change the black shirt, which they put on upon these occasions, until they had avenged the death of the slain man: they also kept his bloody shirt to remind them of their duty. FOSCARINI, Relatione, etc. f. 101. "Li parenti et le donne non si spogliavano mai la camisa nera, che si vestivano subito, se non facevano vendetta, conservando la camisa insanguinita del morto." Among the Maniates, at the present day, "When one of a family is slain, the person who takes upon himself to avenge the injury, often vows not to change his clothes or shave or eat meat, until his revenge is satisfied." LEAKE, Travels in the Moréa, Vol. I. p. 238. A similar custom prevails among the Morlacchi; see FORTIS, Travels into Dalmatia, p. 58: among the Sardinians, see Captain SMYTH's Sketch of the present State of the Island of Sardinia, pp. 142–145: among

Sfakiá, and seems to think it a natural and laudable practice.

Not only were feuds thus perpetuated in families, and transmitted as an inheritance from parent to child, but certain villages were always on ill terms, and sometimes they even made war on one another. The leading hostile hamlets were Gýro and Kámpi, one half of Anópolis siding with Gýro, and the other half with Kámpi. Kalikráti and Askýfo were united, both in their friendship to one another, and in their hatred of Nípros and Asfénto. The strict union which existed between the two latter villages, arose, a long time ago, as follows. A Nipriόte, whose name was Oekonomikó, was shot by an Asfentióte. Most of the Nipriótes were connected with their slain fellow-villager by the ties of blood or marriage, and all of them determined to make a memorable example of the then hated village of Asfénto. They, therefore, attacked the Asfentiótes and actually drove them out of the country, and obtained possession of their houses and property. The poor Asfentiótes went and established themselves at Kofinás in Mesará. "On this account the Nipriótes and Asfentiótes are all relations even to this day, and no other place ever had the force of Nípros and Asfénto."

Whenever any Sfakian who was slain had many relations, his murderer had no chance of impunity, and used always to consult his safety by flight[22]. His house was immediately burnt, and his property taken possession of by the relations of the man he had killed.

among the Druses, see BURCKHARDT, Syria and the Holy Land, p. 202: and among the Afghans, see Mr ELPHINSTONE's Account of the Kingdom of Caubul, p. 166. "Though private revenge is every where preached against by the Moollahs and forbidden by the government, yet it is still lawful and even honourable, in the eyes of the people, to seek that mode of redress. If no opportunity of exercising this right occur, he may defer his revenge for years: it is often transferred from father to son for several generations."

[22] "A Morlack who has killed another of a powerful family, is commonly obliged to save himself by flight, and to keep out of the way for several years." FORTIS, Travels into Dalmatia, p. 58.

It would be easy to fill a volume with instances of the regular exaction of this penalty of blood for blood by the relations of every one who was slain. The usage was by no means confined to Sfakiá, although it seems to have been observed here more invariably than elsewhere in the island, a point which every Sfakian thinks highly honourable to his country. Additional details are however hardly necessary in order to develope the nature of these occurrences. In a country where laws would take the trial and execution of the murderer out of the hands of his accusers, such actions would of course lose the character of honourable vengeance and even stern justice, which they possess in Sfakiá[23], and acquire that of bold and memorable crimes.

These customs remind us of the state of society described in the Iliad and Odyssey, where we find that the murderer either makes compensation to the relations of the man slain by him[24], or goes into exile to avoid their vengeance[25]. Somewhat later there were introduced expiatory rites, which many ancient authors have assigned to the heroic age[26], but which do not seem to have been known to Homer[27]. In ancient times it would every where have been considered a heinous crime, as it would be now in Sfakiá[28], to refuse shelter and protection to such a fugitive.

[23] And which they seem to have had in Scotland only about two centuries ago. See ROBERTSON'S History, and Professor DUGALD STEWART'S, Philosophy of the Active and Moral Powers of Man, Vol. I. Chap. III. 1. (3). p. 186. foll. ed. 1828.

[24] As in HOMER, Il. IX. 632. XVIII. 498.

[25] HOMER, Il. II. 665. XIII. 696. (and XV. 335.) XVI. 573. XXIII. 86. XXIV. 480. Odyss. XIV. 379. XV. 224.

[26] See LOBECK, Aglaophamus, Vol. II. pp. 967–969. (Epimetrum II.)

[27] SCHOLIA VENETA, on Il. XI. 680. Παρ' Ὁμήρῳ οὐκ οἴδαμεν φονέα καθαιρόμενον, ἀλλὰ ἀντιτίνοντα ἢ φυγαδευόμενον. With MUELLER, (Dorien, II. 8. 6. note m. in the Eng. tr. and Aeschylos Eumeniden,) compare LOBECK, Aglaoph. p. 301. See also Mr THIRLWALL'S observations on this subject in his History of Greece, Vol. I. pp. 170–1.

[28] A refusal was never heard of even when the fugitive was a Turk. The fugitives, from all parts of Syria, find an inviolable asylum with the Druses: BURCKHARDT'S Syria and the Holy Land, pp. 203–204.

The existence of such feuds as these, in many Sfakian families, would greatly have paralysed the exertions of the Cretan insurgents against the Mohammedans in 1821. The Oriental Church, however, invites her sons to become "brothers[29];" and the spiritual relationship, thus entered into, is of so solemn and sacred a nature, that, like gossipred, it for ever prevents marriage between those immediately connected with the contracting parties. This religious ceremony was very generally performed, among the Sfakian mountaineers, in 1821, and they were thus enabled, forgiving all mutual enmities, cordially to work together in every attempt to injure their common foe the Mohammedans.

Maniás repeats to me a distich, according to which

> The brothers whom the church doth make
> Are dearer to each other
> Than those who're tied by bonds of blood,
> As children of one mother[30].

The Slavonian ritual, which, I suppose, differs but little from the Greek, also contains "a particular benediction for the solemn union of two male or two female friends in the presence of the congregation[31]." These customs of the mountaineers of Greece and Dalmatia, call to our recollection the old Cretan institutions which sanctioned a close intimacy between those of the same sex[32], and were undoubtedly designed "to revive that generous friendship of the heroic ages, which was so

[29] There is an Ἀκολουθία, or office, for this spiritual fraternization in the Greek Euchology, published by GOAR: (see pp. 898—902.) The custom seems to have prevailed as early as the age of Justinian; (CODINUS, de Orig. C. P. quoted by GOAR, l. c. p. 901.) and although it was forbidden by both Imperial and Ecclesiastical authority, (GOAR, l. c. p. 902.) is still, as we find in Sfakiá, even generally prevalent. Monks, however, were always prohibited from becoming either *σύντεκνοι* or *ἀδελφοποιητοί*.

[30] Καλίτεροι οἱ ἀδερφοὶ τῆς ἐκκλησίας
παρὰ οἱ ἀδερφοὶ τῆς κοιλίας.

[31] FORTIS, Travels in Dalmatia, p. 57.

[32] STRABO, x. p. 483.

celebrated in song, and to add a new motive to the love of glory in the noblest spirits[33]."

Vasíli Khális, a distinguished leader during the war in this island, was once speaking to me, in the presence of old Manúsos, of the wild and savage ferocity of the Sfakians in olden times, and of the readiness with which they shot any one on the slightest provocation. He mentioned a fact which had happened at Thériso, when he was very young. A man was shot by a Sfakian for having killed *two Sfakian dogs*. Hardly had he pronounced the words before old Manúsos said, laughing, "It was I who killed him[34]." The reckless daring of this man's character, his unscrupulous maintenance of what he thought the rights and privileges of the Sfakians, and his having had a numerous body of relations, must have made him one of the most formidable of the wild race to which he belonged. I wonder not that he should have received the tithe of their flocks from the Greeks of the villages of Hághios Ioánnes and Haghía Ruméli, who thus secured to themselves the benefit of his protection. Flocks thus protected were safe from the attacks of every marauder, for if any one of these his ráyas, as he calls them, was ever robbed of a sheep, Manúsos used either to obtain restitution of the stolen property, and ample satisfaction for the offence, "or to shoot the man who had stolen it." Such protection from a powerful Sfakian was so necessary for all those who dwelt near the Rhíza, that several Mohammedans used likewise to obtain it, in the same manner, giving one sheep in ten as black-mail, to secure from molestation the rest of their flocks.

Manúsos's father left him only two sheep: he says that he possessed a thousand when the revolution broke out in 1821. This is hardly surprising.

33 Thirlwall, History of Greece, Vol. I. end of ch. VII. p. 290.

34 Ἐγὼ τὸν ἐσκότωσα.

Whenever a married woman was *suspected* of faithlessness, or a single one of frailty, her hours were from that moment numbered, and her end was so tragical and shocking to all the feelings of natural affection, and even to the ordinary notions of humanity, that one can hardly believe such a practice to have been observed on the very confines of civilized Europe, and in the nineteenth century, by any Christian people. Her nearest relations were at once her accusers, her judges, and her executioners.

I will mention, in the very words of my informant, a single instance of the ruthless severity with which they punished such an offence. Thirty-five years ago a young wife, the cousin of Manúsos, who was present at the recital, as he was also at the events described, was suspected of having broken her marriage-vow. The charge was not proved, but the suspicion became general, and her father at once consented to leave it to her near relations to decide as they thought best respecting her. Their decision was soon taken, and all those of them whose dwellings were at Askýfo went thence to Anópolis, where their destined victim lived, in order that they might assist in executing the sentence pronounced against her, and of which she was as yet totally ignorant. They went, to the number of between thirty and forty, to her home, seized her, and, after tying her to a tree, made her person the mark at which all their muskets were pointed and discharged. It happened that though more than thirty balls lodged in her body, the next moment her bosom heaved, and she still breathed. One of her executioners immediately drew his pistol from his girdle, placed it close to her breast, and fired. After this she breathed no more.

The suspected partner of her guilt was not shot "for he was of a powerful family;" but the Protópapas[35],

[35] Τὸν ἀφόρισεν ὁ Πρωτόπαπας.

the unfortunate woman's father, excommunicated him, and, in consequence of the sacerdotal ban, he not only perished himself by falling over a precipice, but all his brothers likewise came to untimely ends.

Among the Cretans of ancient times, the adulterer was condemned to the ignominious punishment of being crowned with wool in the presence of the magistrates of the city[36].

[36] AELIAN, V. H. XII. 12. Ὅτι ἐν Κρήτῃ, ἐν Γορτύνῃ μοιχὸς ἁλοὺς ἤγετο ἐπὶ τὰς ἀρχάς, καὶ ἐστεφανοῦτο ἐρίῳ ἐλεγχθείς· τὸ δὲ στεφάνωμα κατηγόρει αὐτοῦ, ὅτι ἄνανδρός ἐστι καὶ γύννις καὶ εἰς γυναῖκας καλός· He was also fined, καὶ ἀτιμότατος ἦν, καὶ οὐδενός οἱ μετῆν τῶν κοινῶν.

CRETAN MOUNTAINEER OF THE SIXTEENTH CENTURY.

CHAPTER XXXVIII.

ANCIENT AND MODERN COSTUME OF THE CRETANS. DEPARTURE FROM HAGHIOS DHEMETRIOS. ARADHENA. MULES ALONE FIT TO TRAVERSE THE STONY PATHS OF SFAKIA. STEEP AND TEDIOUS DESCENT TO THE SHORE. FOUNTAIN OF ST PAUL. TRADITIONS OF OTHER FOUNTAINS PRODUCED BY MIRACULOUS POWER. CRETE FREED FROM ALL NOXIOUS ANIMALS BY HERACLES OR ST PAUL. ST ISIDORE'S MIRACLE AT CHIOS. ARRIVAL AT THE VILLAGE OF HAGHIA RUMELI. THE SITES OF TARRHA AND POEKILASSOS.

Fashions change but little in the east, and the dress of the Cretan peasant is very interesting, even at the present day, from its resemblance to that worn by his ancestors more than two thousand years ago. If we view the Cretan costume in the sixteenth century, we shall find that it still more closely resembles that of the ancient inhabitants of the island. In the engraving on the previous page we notice that the beard, the bow and quiver, the sword and long knife[1], were all preserved. The χιτών, *tunica* or shirt, according as we give the garment its Hellenic, its Roman, or its English name, was still exposed in all its amplitude to the eye: and Galen's description[2] of the ὑποδήματα or boots, worn by the Cretans of his time, is a sufficiently accurate account, both of those described by Foscarini[3], and of those which I find still preserving their ancient form and name, and alone worn by the rural population in every part of the island.

Thevet, who visited the island near the middle of the sixteenth century, says, that to cut off the beard was as great an indignity, in the opinion of every Cretan, as to

[1] The long sword at his side, and a knife in his girdle, were parts of every Sfakian's dress till the year 1830.

[2] Galen, Comment. in Hippocrat. de Artic. iv. 14. Vol. xviii. p. 682. ed. Kuehn. Ἔτι καὶ νῦν οὗτος ἀσκεῖται κατὰ τὴν ἡμετέραν Ἀσίαν, καὶ κατὰ τὴν Κρήτην οὐχ ἥκιστα, μέχρι κνήμης μέσης ἀνήκων. ἔστι δ' ἱκανῶς ὑπ' αὐτοῦ πολυσχιδὲς τὸ δέρμα καθ' ἑκάτερον τὸ μέρος ἀρίστερόν τε καὶ δεξιὸν ἴσον ἐντεταμένον τε καὶ ἰδίᾳ συντετρημένον ἐπὶ τοῖς πέρασιν, ὡς ἱμάντι διεκβαλλομένων τῶν τρημάτων συνάγεσθαι κατ' ἄλληλα τὰ τεταγμένα τοῦ δέρματος τμήματα—χρῶνται δὲ αὐτῷ μάλιστα παρ' ἡμῖν οἱ κυνηγέται θηρεύοντες, ἐπειδή περ ἐν τραχέσι καὶ ἀνωμάλοις χωρίοις ἀναγκάζονται πολλάκις οὐ τρέχειν μόνον, ἀλλὰ πηδᾷν. ἡ δ' αὐτὴ καὶ τοὺς Κρῆτας αἰτία τοιούτοις ὑποδήμασι χρῆσθαι ἔπεισε, διά τε τὰ μεγέθη τῶν ὀρῶν τῶν ἐν τῇ νήσῳ καὶ τὰς τραχύτητας.

[3] Foscarini, Relatione etc. "Li assomiglio quasi alli selvatichi Irlandesi, perchè, come quelli nell' abito, et nell' aspetto, et nelle armi, sono diversi dalli altri; cosi questi, con li capelli et *barbe longhe*, *con stivali alti che si ligono alla cintura*, *che mai li escono di gamba*, *portando la camisa che li pende davanti et da driedo*, *con un gran pugnal davanti*, *et la spada* all' usanza greca vanno." The two words used by the Cretan to denote his boots are the ancient ὑποδήματα, and στιβάνια, derived from the Italian *stivale*.

cut off the nose or ears. "Ilz ont les longues barbes en si grande recommandation que les contraindre à icelles raire, ou copper, leur est vn supplice et ignominie, non moindre que si on leur faisoit copper le né, ou les oreilles. Et de fait pour lors que i' estois en ladite ville, ie vizi condemner vn Candiot, qui auoit dōné vn coup de flesche à vn autre, à auoir la barbe abbatue et raise en place publique[4]."

Not only was the bow common in the sixteenth century[5], but wounds from arrows are frequently mentioned in the details of the siege of Candia, which took place only a hundred and fifty-five years ago. In speaking of the Sfakians, Foscarini says (in 1596), that they always carried their bow and quiver, and were most excellent archers: "Vanno sempre con l' arco sulle spalle et al fianco con un carcasso pieno di frezze, tirate da loro eccellentissimamente, come molti anco sono eccellentissimi archibusieri." Garzoni also mentions the difficulties encountered by the Venetians when they attempted to replace the ancient bow by a modern innovation[6].

Another article of dress sometimes used by the Cretans is a very short cloak, falling only a little below the girdle. This I suppose to resemble the *Kretikon*, mentioned by Eupolis[7] and Aristophanes[8], and stated by

[4] THEVET, Cosmographie de Levant, Chap. VIII. fol. 26.

[5] BELON, Singularitez etc. Fol. 6. "Ce n'est pas à tort que les Cretes furent anciennement dediez à Diane: car encore pour le iour d'huy suyuant ceste antiquité, s'adonnent par vn instinct naturel, et des leur enfance, à tirer l'arc Scythique: et mesmement vn petit enfant du berseau courroussé et pleurant, s'appaise en luy monstrant seulement vn arc, ou luy baillant vne flesche en la main." The expression "l'arc Scythique" calls to mind the passage of DIODORUS SICULUS, V. Ἀφ' ἧς αἰτίας μάλιστα παρὰ τοῖς Κρησὶν ἐζηλῶσθαι τὴν τοξικήν, καὶ τὸ τόξον Σκυθικὸν ὀνομασθῆναι. On the Cretan archers, see the passages of ancient authors collected by MEURSIUS, Creta, pp. 177—180.

[6] See below, Appendix I.

[7] EUPOLIS, in Photius, Lex. Vol. I. p. 178.

Οὐ πάνυ ταχύ
ῥίψας ἐμοὶ τοῦτ', ἀναβαλεῖ τὸ Κρητικόν;

[8] ARISTOPHANES, Thesm. 730.

Σὺ δὲ τὸ Κρητικόν
ἀπόδυθι ταχέως.

CRETANS IN THEIR CLOAKS.

some lexicographers[9] to have been a short light cloak, just as much as the modern boots resemble the ancient throughout in the island.

It is worthy of remark, that the Sfakian cloak agrees less than that of the other Cretans with the descriptions of the *Kretikon*. The cloak of these mountaineers is large and heavy, and is of its natural whiteness, while those of the rest of Crete are dyed[10], as the Kretikon used to be[11].

It is the Sfakians alone who wear their cloaks through any considerable part of the year: the other Cretans only make use of them a little in the winter months. The ancient Cretan seems to have worn his cloak all the year round[12], just as the modern Albanian puts on his shaggy white capote to defend himself as well from the heat of July and August, as from the cold of December and January.

Little as is the change which has taken place in the fashion of the Cretan's dress, there is yet a considerable difference in its material in ancient times, and at the present day. Both the under and upper garments of the Dorians were woollen; and the same dress was adopted by the Athenians, and had become general in the time of Thucydides[13]. Thus the linen khiton or shirt, among the ancients, was thought a luxury: and yet at the present day the Sfakian mountaineer frequently wears a silken shirt, although probably it is only washed for

[9] PHOTIUS, l. c. *Κρητικόν· βραχὺ ἱμάτιον.* HESYCHIUS, *Κρητικόν· ἱματίδιον λεπτὸν καὶ βραχύ, τὰ γὰρ τοιαῦτα, Κρητικὰ ἔλεγον.*

[10] They are of a brown colour.

[11] PLINY, XXVI. 66, in speaking of the phycos thalassion, says, "Tria autem genera ejus: latum, et alterum longius, quadamtenus rubens: tertium crispis foliis, quo in Creta vestes tinguunt.

[12] STRABO, X. p. 483. *Χαμαὶ δὲ καθήμενοι διαιτῶνται μετ' ἀλλήλων ἐν φαύλοις τριβωνίοις, καὶ χειμῶνος καὶ θέρους τὰ αὐτά.* HERACLIDES, de polit. (in Gron. Thes. Vol. VI. p. 2824, quoted by MEURSIUS, Creta, p. 184. and by CASAUBON, on Strabo, l. c.) *Οἱ ἐν Κρήτῃ μετ' ἀλλήλων διαιτῶνται ἐν ἑνὶ ἱματίῳ θέρους καὶ χειμῶνος.*

[13] THUCYDIDES, I.

Easter Sunday during the whole time while it is in use. It is occasionally taken off, like that of an Albanian, in order that its owner may free him from some of those active little animals which have long enjoyed an undisturbed shelter within its hospitable folds[14].

May 6, continued.

On leaving Hághios Dhemétrios, we crossed the low ridge which bounds the plain of Anópolis in this direction, and in about half an hour arrived at the edge of a chasm running south of the village of Arádhena. It commences to our right near the foot of the loftiest ridge of the Sfakian mountains, and extends down to the sea. It is no easy matter to cross this nearly perpendicular cleft in the solid rock, forming as it does a narrow ravine of several hundred feet in depth, down and up the steep sides of which we have to pass. The road, if road it can be called that road is none, winds along each side, changing its course every ten or twelve paces. These turns were the only very dangerous points: at these I more than once expected to see my horse miss his footing, for the stones on which he had to tread were large and slippery, and more suitable for goats than for horses to step on. The descent and subsequent ascent, on the opposite side, to the village of Arádhena, occupied about twenty-five minutes.

Accounts which I had heard at Anópolis, of ancient remains existing at Arádhena, had led me to expect that I should discover here clearer and greater vestiges of ancient power and prosperity than I had found yesterday at Anópolis; but here, as elsewhere, rude realities destroyed my hopes.

Some slight foundations of the walls of ancient buildings; a few tombs, in one of which several golden

[14] Captain Maniás once told me, that during the war he used always to have two shirts with him, one besides that on his back, in order to be able to change. "Pray how often did you change your shirt then?" "Ὅταν ἔγενε ψυλλισμένο," was his reply.

ornaments were found about a year ago; and a sepulchral bas-relief, the execution of which is not very remarkable, are the only remains which serve to indicate Arádhena or its immediate neighbourhood as the site of an ancient Cretan city.

The mention of an ancient city called Aradena, along with Anópolis and Port Phoenix in the Synecdemus of Hierocles, seems to point plainly to Lutró as the site of the last-named city; and therefore I will leave the consideration of the other two cities until my arrival at Lutró, on returning from the almost inaccessible fastnesses of Haghía Ruméli, and Samaría.

The modern village of Arádhena contains only thirty-eight houses: the ground capable of cultivation round about it is not very extensive, but is of a decidedly better quality than that of the plain of Anópolis, as is evident from the good crops of barley which now cover it. The chief produce of the village is corn: its silk does not exceed a few okes yearly. The Greek in whose house I took up my quarters, for the short time which I remained in the village, is an Ionian, that is, a native of one of the islands of the Septinsular Republic, and is therefore entitled to British protection. He is thus exempted from the payment of the kharatj, to which all rayas, both of the Porte and of Mehmét-Alí, are still subject. He married a Sfakian woman of this village, having first joined the Sfakians at the outbreaking of the revolution. His long intercourse with them has communicated to him a good many Sfakian modes of expression, although the Italianisms of his native language are still visible through this thin covering.

Some ancient tombs have lately been discovered and opened about three quarters of a mile to the west of Arádhena. I examined one of them: it is six feet wide and ten feet long, and has a niche in each of its four sides.

I left Arádhena at four o'clock, and soon had the village of Livadhianá about a mile and a half on my left, a little to the west of the chasm which we had crossed before reaching Arádhena. The slope of the mountain is here covered both above and below us with *pévki.* We do not proceed a mile before we fall in with some of Maniás's innumerable Sýntekni and cousins at a sheep-cote. Meat, new cheese, and wine are produced and partaken of by my talkative and amusing guide, as if he was determined to make up, in this single week, for the abstinence of the whole Lent from which he has just escaped.

Soon after this our course changed to the south-west, and we began to approach nearer to the shore. At a quarter past five, Sélino-kastéli became visible. The paths which we followed were over rugged rocks, and none of them, I should think, had *ever* received any labour from the hand of man. Here my horse, though very sure-footed, fell, and made two unsuccessful attempts to rise, each of them followed by a fresh fall, before he recovered himself. On dismounting, to see what hurts he had received, I was surprised to find that they consisted only in some flesh wounds, which bled considerably, inflicted by sharp edges of the rock, on two of his legs.

The only animal which one can ride with pleasure in these regions is a mule. The mere size of a horse's hoof makes it a difficult matter for him to pick his steps in many places; and the very art of picking his steps is but little known to any horse, while every mule is a proficient in it.

Soon after my horse's roll, and my sufficiently narrow escape from what would have been, at least, a very disagreeable fall, we arrived at the summit of a precipice, the zigzag descent down the face of which is so steep, rocky, and dangerous, that I was greatly disposed to send Maniás back with my horse, and to proceed with his mule. It was now, however, late; and, since I might perhaps be benighted before reaching Haghía Ruméli, and was also totally unacquainted with the wild region before me, I determined to attempt to get down with the two steeds. This zigzag road winds along the face of a rocky and almost perpendicular precipice of very considerable elevation. Had my horse missed his footing at any one of the turns of the zigzag path, it must have been his last trip. After getting down this steep and difficult descent in safety, we were still at a great elevation above the level of the sea. Although the side of the mountain now slopes off at a much smaller angle of inclination to the horizon, yet it is still so steep that the path continues to be of the zigzag kind nearly all the way down: the face of the mountain here presents a loose surface of stones and a little soil, along which we continued to wind to and fro for about an hour and a half.

At length we arrived near the sea-shore, somewhat to the west of the commencement of our descent over the precipice; and going nearly due west for above a mile, we reached the little chapel of Hághio Pávlo. Close by it a most plenteous spring of fresh water rushes out of the beach, and, forming a rapid streamlet, flows for a few paces before it loses itself in the sea. The Florentine traveller in Crete, Buondelmonti, whom I have so often referred to, speaks of this streamlet as being half the size of his own Arno[15]: a most extravagant exaggeration, as every one must admit who visits St Paul's spring after having been

Là dove l'Arno signoril cammina.

Another fountain in the island is also honoured by the derivation of its name from St Paul, who is said to have used its water to baptize the Cretan converts. An English traveller went, about a hundred and fifty

15 Buondelmonti, in Cornel. Cret. Sac. I. p. 25. and p. 85. He speaks of the lofty cypresses which grow on these rocks of the southern coast, and adds, "Medioque ipsarum, per longum iter Sancti Pauli ecclesiam vidi prope, cujus tam eructatione maximam ex paucis foraminibus frigidissimae aquae exultat, *quia* vix in duplum noster adsumit Arnus, que per modicum spacium in mare prorumpit."

years ago, to see a cave, ten miles to the east of Hierápetra, "where they say St Paul preached: it is a large chapel having twelve pillars all cut out of the rock, which was done by the Christians in the night-time. Close by is a fountain, where they say he used to baptize, and it is now called St Paul's fountain: the water thereof is very good to cure such as have sore eyes[16]."

We may best compare with this Sfakian fountain of St Paul, another of Dionysius, in the Strophades, which is still a living and perennial proof of that saint's miraculous powers. "Its fresh and wholesome waters ooze from the rocks at about fifty yards from the sea, and preserve their purity almost on the brink of the briny element[17]." Another copious source, in Epirus, is assigned to the exertion of miraculous power by St Donatus[18]. Fresh springs, similar to this, are also found close to the sea in Ithaca, and in the neighbourhood of Mount Tor, as well as along the whole shore at the head of the Elanitic gulf near Akaba[19].

A Swiss traveller, who remained about a week in Crete, when on his way to Jerusalem, in the year 1542, speaks of a similar fountain of fresh water, seen close to the shore at Megálo-Kástron, and which had lost its saline nature in consequence of having been blessed by Saint Francis[20].

St Paul is said to have conferred other and even still greater blessings than these springs on the Cretan people, when he visited them after his two years residence at Rome: he freed their island from all wild

[16] RANDOLPH, State of Candia, p. 77. ed. 1687.

[17] WADDINGTON, Letter from the Strophades, appended to the Present Condition of the Greek Church, p. 199.

[18] LEAKE, Travels in Northern Greece, Vol. IV. pp. 64–65.

[19] LEON DE LABORDE, Arabia Petraea.

[20] IODOCUS A MEGGEN, Peregrinatio Hierosolymitana, Dilingae, 1580. cap. V. p. 56. "Huic loco puteus est vicinus sane nobilis, cujus aquam ob vicinum mare olim salsam, post S. Francisci benedictione et prece in dulcem ac potabilem conversam aiunt."

beasts and noxious animals. Unfortunately for the Christian legend, we find that Crete was just as much commended in ancient times[21] as it is by its present inhabitants[22], on account of its freedom from all such animals. The pious heathen used to ascribe this blessing to the beneficent exertions of Heracles[23]: as the Christian attributes it to St Paul[24]. The Cretans, however, bring forward no such conclusive proof of the *destruction* of serpents here, as is vaunted by the inhabitants of Malta, in the so-called *lingue di San Paolo*, which they find imbedded in the common stone throughout their island[25].

[21] AELIAN, N. A. III. 32. Ἡ Κρήτη τοῖς λύκοις καὶ τοῖς ἑρπετοῖς θηρίοις ἐχθίστη ἐστίν. PLUTARCH, de capienda ex inim. util. p. 86. b. Ἄθηρον ὥςπερ ἱστοροῦσι τὴν Κρήτην. GREGORY NAZIANZEN, Carm. XIII. 49.

Χώρην μέν τιν' ἄθηρον ἀκούομεν, ὥς ποτε Κρήτη.

PLINY, N. H. VIII. 83. "In Olympo Macedoniae monte non sunt lupi: nec in Creta insula. Ibi quidem non vulpes ursive, atque omnino nullum maleficum animal praeter phalangium." See also SOLINUS, c. XVII. and other passages in MEURSIUS, Creta, p. 101. or WYTTENBACH on Plutarch, l. c. Tom. VI. p. 614.

[22] A living Cretan poet, PSARUDHAKES, expresses the well-known boast of his fellow-countrymen of the present day, when he says:

Ἄχ! πατρίς μου ποιὸν καλὸν
δὲν ὑπάρχει εἰς ἐσένα;
σὺ οὐδὲ φαρμακερὸν ζῶον
δὲν γεννᾷς κανένα!

My native land! what thing that's good
Is not possessed by thee?
No noxious animal, in Crete,
Can any one e'er see.

[23] DIODORUS SICULUS, IV. 17. Τιμηθεὶς (ὁ Ἡρακλῆς) ὑπὸ τῶν ἐγχωρίων μεγαλοπρεπῶς, καὶ βουλόμενος τοῖς Κρησὶ χαρίσασθαι, καθαρὰν ἐποίησε τὴν νῆσον τῶν θηρίων· διόπερ ἐν τοῖς ὕστερον χρόνοις οὐδὲν ἔτι τῶν ἀγρίων ζῴων ὑπῆρχεν ἐν τῇ νήσῳ, οἷον ἄρκτων, λύκων, ὄφεων ἢ τῶν ἄλλων τῶν τοιούτων. Zeus's birth in the island is assigned as the cause of this freedom from wild and noxious animals, by the author of the book de Mirab. Auscult. c. 84. Compare AELIAN, N.A. v. 2.

[24] RANDOLPH, The present state of the Islands in the Archipelago, p. 93. ed. Oxford, 1687.

[25] Every traveller on the *continent* of Greece sees serpents very frequently: thus the stork becomes a valuable friend of man, and, in Thessaly, it was anciently a capital offence to kill the bird, as it has since been in other countries: see the book de Mirab. Auscult. 23. Since the revolution they are said

Doubtless the destroyer of all these noxious animals, whether Heracles or St Paul, was entitled to the gratitude of the Cretans. But still, according to traditions, which undoubtedly equally deserve our attention, other places have received even more signal proofs of favour from other canonized individuals. At Chios, St Isidore converted, at the time of his martyrdom, a useless *liquid* which, till that time, had been constantly dissipated by the rain, as it oozed from the tree which produced it, into *solid* gum mastick, which became the staple produce, and one of the main sources of wealth and prosperity of the island[26].

On leaving the spring and chapel of St Paul, about seven o'clock, we follow the shore, but the shades of evening spread around us, and I see but indistinctly the grand features of the scenery by which we are surrounded. At twenty minutes before eight, we reach the entrance of the valley of Haghía Ruméli, and Samaría. The bold hanging mountains on each side of the glen, and the noise of its river as it rolls along its rocky bed, make me impatient to see, in broad daylight, what I suppose must be the grandest and at the same time the most picturesque spot in the island. A mile's ride up the glen brings us to the village of Haghía Ruméli. Notwithstanding the lateness of the hour, I got very good lodgings in the cottage of one of the villagers, to whom Maniás, who had never before penetrated into this innermost and most inaccessible fastness

said to have disappeared from many parts of the *kingdom* of Greece, by the emancipated inhabitants of which they were considered as *Turkish birds*, for the humanity and affection of the Mohammedans had been bestowed on them even more remarkably than on the rest of the feathered tribe. In Crete I never saw a serpent during my whole stay in the island: they are, however, sometimes found. See WESSELING, on Diod. Sic. l. c. and SIEBER, Reise nach der Insel Kreta, Vol. II. p. 99.

[26] NICOLAUS PEPAGOMENUS, in Leo Allatius, de quorund. Graecor. Opinat. p. 176. *Οἱ τοῦ μάρτυρος ἆθλοι, καὶ τὰ λαμπρὰ κατὰ τῶν ἐχθρῶν ἀγωνίσματα—εἰς ὃ νῦν ὁρᾶται πρὸς λίθου φύσιν ἐγγὺς μετεποίησεν, ἅμα τε τούτου τοσαύτην εἰργάσατο τὴν ἐπίδοσιν, ὡς* &c.

of Sfakiá, was known by name, and who made me extremely welcome.

May 7.

The villagers point with exultation to the lofty mountains by which they are surrounded, and say that they are the best fortress to be found in all Crete, and the only place within which the Turks never penetrated, during the long war between 1821 and 1830. On walking through the village I find some houses in ruins; and learn, from old people, that they were destroyed by the Mohammedans when they invaded Sfakiá in 1770. Thus the greater part of a century suffices not, in the Ottoman dominions, to repair the destruction caused by a month's warfare!

My hostess's beautiful unmarried daughter, on my asking her whether she had ever been at Khaniá, exclaimed, "God forbid!" and assigned as a reason for not wishing ever to leave her native glen, that she supposes the cottages of Haghía Ruméli appear more beautiful to her now, than they would after she had seen the fine houses of the city.

I learn that at a place called Trypeté[27], near Vukoliási[28], between Haghía Ruméli and Súia, and somewhat nearer to the latter place than to the former, at a spot the road to which from here is so bad that the place may be called inaccessible, except by sea, are found Hellenic remains, situated about a mile from the shore. The hill on which they stand is visible from the sea. It will be very easy to determine the name of the ancient site of which these are remains, if we can succeed in fixing a city on the shore below Haghía Ruméli.

Tarrha, which is interesting as one of the earliest localities of the Apollo-worship, has been supposed, by some very distinguished scholars, to have been inland;

27 In the island of Mélos also is a village Trypeté, which derives its name "from the small catacombs with which the hill is pierced in every part." LEAKE, Travels in Northern Greece, Vol. III. p. 81.

28 Εἰς τὴν Τρυπητή, εἰς τὸ Βουκολιάσι.

but I think there is little or no doubt, that its site is on the shore, at the very entrance of this glen of Haghía Ruméli[29].

On examining the spot I found very slight vestiges of antiquity, but the Tuscan priest Buondelmonti describes considerable remains of a temple and other buildings at the site of the ancient city, as existing when he was in Crete, more than four hundred years ago[30].

If Tarrha was at the entrance of this glen, as I think it clearly must have been, it follows that Trypeté is the site of Poekilassos[31].

Perhaps the inscription, of which Buondelmonti speaks[32], may be partly owing to a passage which he had read in Solinus[33].

These words, "aedem numinis, praeterquam nudus vestigia, nullus licito ingreditur," are remarkable, not on account of the singularity of the usage described by them, but, as Lobeck has somewhere observed in his Aglaophamus, on account of the undoubted generality of the practice in question. A similar feeling prevailed among the Jews[34], and with the Pythagoreans[35]. At

[29] The chief authority is the STADIASMUS, p. 496. ed Gail. Ἀπὸ Φοίνικος εἰς Τάῤῥον στάδ· ξ′. πόλις μικρά ἐστιν· ἔχει ὅρμον. Ἀπὸ Τάῤῥου εἰς Ποικιλασσὸν στάδ· ξ′. πόλις ἐστί, καὶ ὅρμον ἔχει, καὶ ὕδωρ.

[30] BUONDELMONTI in Cornel. Cret. Sacr. Tom. I. p. 85. "Portum antiquae urbis desolatum—hodie Romelum—ingentem aspicio templum in subversione volutum, quo omnia genera marmorum—sine ordine jacent. Aspicio idolorum sine capitibus busta, et ab altero latere templi caput Veneris vel Dianae inveni, quod super omnia pulcherrimum videbatur.—Columnae cisternaeque cum hedificiis amplis omnia patescunt."

[31] STADIASMUS, l. c.

[32] BUONDELMONTI, l. c. "Cernimus nimis deletas Graecas literas scriptas, sequimur deinde eas, heu scissum erat, sed per conjecturas sic cernere potui: Munda pedem, vela caput, et intra."

[33] SOLINUS, cap. XI. "Cretes Dianam religiosissime venerantur, Britomarten gentiliter nominantes, quod sermone nostro sonat virginem dulcem. Aedem numinis praeterquam nudus vestigia nullus licito ingreditur. Ea aedes ostentat manus Daedali."

[34] EXOD. III. 5. JOSH. V. 15. ACTS, VII. 33.

[35] IAMBLICHUS, V. P. §. 85. p. 182. and §. 105. p. 228. Θύειν χρὴ ἀνυπόδετον—and Ἀνυπόδητος θῦε καὶ προσκύνει.

A. Schranz del. L. Haghe Lith.

ENTRANCE OF GLEN OF HAGHIA RUMELI.

the present day, an oriental bestows the same mark of respect on every equal or superior in whose presence he may find himself, so that to take off the shoes in the east has almost the same meaning as to take off the hat has with us.

The only Christian nation by whom, as far as I know, the ancient oriental and pagan custom, of entering the house of prayer barefoot, is still practised, is that of the Armenians[36]; who also prostrate themselves to the ground in their adorations, much in the same manner as is done by the Mohammedans[37].

[36] Smith and Dwight, Missionary Researches in Armenia, Letter vii. p. 140.

[37] The Armenian system of religion resembles the Greek in many of its absurdities. One of their authors defends the worship of "the God-bearing cross," on the ground that it is offered "not to the visible matter, but to the invisible God who is in it." More scrupulous, however, in some things, than the Greek, the Armenian extends not his profane iconolatry to the pictures of the saints, but holds that, "as neither a name nor an image, without the substance, is to be worshipped, and as the saints are not every where present to dwell in their images, *as Christ is in his*, their images are not to be worshipped." Nerses Shnorhali, Unthanragan, pp. 132–133. quoted by Smith and Dwight, l. c. Letter vii. p. 142.

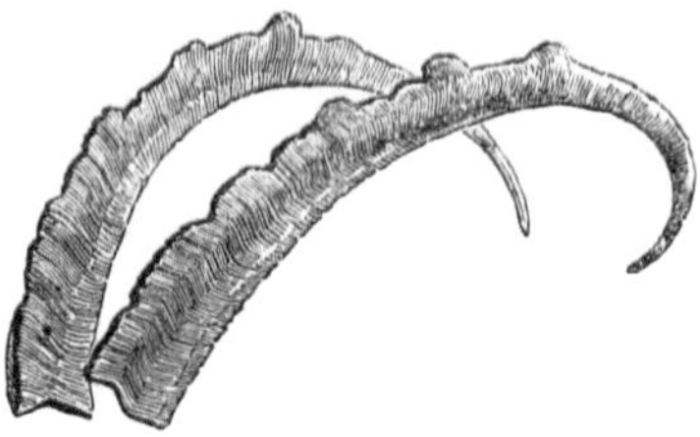

HORNS OF THE CRETAN WILD-GOAT.

CHAPTER XXXIX.

HAGHIA RUMELI TO SAMARIA. DIFFICULTY OF PASSING, EVEN WITH A MULE, ALONG THE GLEN. STRIKING PASS CALLED "THE GATES." KEFALOVRYSIS. A SITE CALLED "THE REFUGE OF THE HELLENES." CAENO. LOVES OF APOLLO AND ACACALLIS. CYPRESSES IN THE GLEN OF HAGHIA RUMELI. THE CRETAN AGRIMI IS THE WILD-GOAT, AND NOT THE BOUQUETIN OR IBEX. DEPARTURE FROM SAMARIA.

May 7, continued.

At half-past eight I left this sequestered village, intending to explore the inner recesses of the glen as far as Samaría. The path in some parts is so narrow, where it winds round the abrupt precipices, that no horse would be able to pass along it. At one spot it was necessary to take off the saddle from the mule on which I was mounted, during the greater part of the way, in order to avoid the likelihood of precipitating the animal into the river below.

In the first half hour after leaving Haghía Ruméli, I crossed the river five or six times, and then arrived

A. Schranz del. L. Haghe Lith.

PASS OF HAGHIA RUMELI.

at a very striking pass, represented in the opposite plate, and called "The Gates[1]."

The width of this lofty chasm is about ten feet at the ground, and widens to about thirty, or at the most forty feet at the top. The length of way along which we have to pass in the middle of the rapid stream, is about sixty paces: and for a hundred paces further we are more in the water than out of it, having to cross the torrent several times. After gazing awhile on the grandeur of this defile, I left it at half-past nine. In about twenty minutes more the rocks again contract, so as to become nearly perpendicular. I observe several pieces of wood in the river, which elicit an observation from my Ruméliόte guide[2]. A little before ten we reach a spot called "the Turk's pass[3]," from the fact of a Mohammedan having been killed there in 1770.

At twenty minutes past ten we arrive at a cluster of fine plane trees, and a copious source called Kefalovrýsis[4], which now supplies the river with half its water; and which, after the melting of all the snow on the mountains above Samaría, towards the end of July or the beginning of August, still causes a considerable stream to flow from this spot to the sea below Haghía Ruméli. The contortions of the rocks in this neighbourhood shew how violent must have been the operation of the causes which threw them into their present shapes.

Cypresses may be noticed scattered over the mountains of this glen along its whole extent; and as we approach Samaría, they are seen in great numbers. We arrived in the village at half-past eleven.

I find the belief in the ancient site said to exist above Samaría, and to have been the last refuge of the ancient Hellenes[5], is entertained by the Samarióte peasant

[1] Τζὴ Πόρταις.

[2] He said, Οὖρα τοῦτα τὰ ξύρα φέρει ὁ πόταμος ἀπὸ μέσα: where οὖρα is used for οὖλα or ὅλα, and ξύρα for ξύλα.

[3] Εἰς τοῦ Τούρκου τὸ πέραμα.

[4] Κεφαλοβρύσις.

[5] Compare what is said above, p. 85.

who undertakes to shew me the way to them, and by most of his fellow-villagers. After reposing for a few minutes, under the shade of an overhanging rock, I commenced the ascent of the lofty and very steep mountain, on which the ancient remains were said to exist, after first winding round the side of an adjacent mountain, to the south-west of the village, and which is thickly covered with pines and cypresses. A good deal of snow is still lying on all the neighbouring mountains.

A steep and tiresome ascent of a full hour brought us to the alleged site of the ancient city. The existing vestiges are remains of walls, constructed of irregularly shaped stones, most of which are small, and which were united by cement. These remains are five or six feet high where they are best preserved, and may be traced for a considerable distance round the mountain.

Just above this wall are seen vestiges of two rooms, each about ten paces square: the wall which divided them is standing to the height of about three feet; and that on the western side is inserted as it were in the side of the mountain, and is ten feet high.

About a hundred paces above these slight ruins, is seen what I suppose once to have been a cistern, although it is not covered over with cement in its interior. Its length is eight paces, and its width three paces and a half. It is partly filled up, its present depth not being above five or six feet.

As to Hellenic remains my ascent ended in disappointment; but I was most amply repaid for my labour by the loveliness of the day, and the wildness and magnificence of the scenery around me. On my throwing out some slight doubts about the vestiges which I saw being *very* ancient, and suggesting that they could hardly belong to the celebrated "Hellenes," my Samarióte guide exclaimed, in the tone of one half offended at my ignorance or incredulity: "*Here* was the end of them, my good Sir[6]!" as if the matter had been one on which

[6] Ἐπὰ ἐτελειώσανι καλέ.

his local information entitled him to pronounce with authority.

The Rumeliốte, who, as well as Maniás, accompanied me from Haghía Ruméli, has served as a sailor on board several vessels, and has thus sadly corrupted the purity of his Sfakian Greek. Some of his expressions were hardly intelligible to the Samariốte[7].

I descended into the valley at a distance of about three quarters of a mile from Samaría, on the way to the Xylóskalo, and did not arrive at the village till nearly three o'clock.

Towards the foot of the mountains, to the northwest, in the direction of the Xylóskalo, and about three miles from Samaría, is the monastery of Hághios Nikólaos. It is surrounded by the largest cypresses in the island, as all my three Sfakian companions agree in assuring me. They tell me many stories of these trees, which they would regard it sacrilege to cut; and the events which they believe to have happened, when one of them was felled by an impious hand many years ago, shew that the credulity of these brave and simple mountaineers is full as great as it was when the wonder-working Saint came from the Holy Mountain to visit them, awhile before the out-breaking of the Greek revolution.

The account given, by my guide, of the venerable cypresses, which surround the church of Hághios Nikólaos, reminded me of the Roman poet's description:

> Est urbe egressis tumulus templumque vetustum
> Antiquae Cereris; juxtaque antiqua cypressus,
> Religione patrum multos servata per annos.

The Graces, who were peculiarly an object of religious worship in ancient Crete[8], were identified, as we have already seen, with these beautiful trees[9].

7 He said "στὴν πορτέλλα," whereon the Samariốte asked him "στὸ παραθύρι ἤτονι;"

8 Apollodorus, III. 15. 7. and Heyne, p. 881. Compare Pausanias, IX. p. 780. and Siebelis, I. p. 17.

9 Vol. I. p. 252. It would be easy to give additional instances of the comparison

On the cypresses of Crete in general, and on those of the neighbourhood of Tarrha in particular, it is sufficient to refer to what has been already said[10], and to the observations made by Bentley in his letter to Mill[11].

The ancient legend of Apollo's purification by Carmanor, at Tarrha[12], gives us the name of a city which must have existed among the mountains in this neighbourhood. Carmanor had a grand-daughter Carma, and the Cretan goddess Britomartis, Artemis's favourite companion, was the offspring of Zeus and this damsel. Now the birth-place of Britomartis is said to have been Caeno[13], and it is therefore probable that Caeno was situated somewhere hereabouts, either on this so-called refuge of the Hellenes, or near Hághios Nikólaos; and that the mount Carma mentioned by Pliny[14], was in the immediate neighbourhood of the town.

I have already spoken of the loves of Apollo and Acacallis[15], the daughter of Carmanor: and we now behold their supposed locality.

> Oft in glimmering bow'rs and glades
> He met her, and in secret shades
> Of woody *Tarrha's* inmost grove.

comparison of beautiful women with the cypress-tree there mentioned. The following are Sfakian distichs:

> Τὸ κυπαρίσσι ῥέγομεν
> τὸ μυρισμένο ξύρο,
> ἀποῦ σου μοιάζει, μάθια μου,
> στὸ μάκρος καὶ στὸ ψήρο.

The last word is ψήλος· εἰς τὴν ὑψηλότητα is the sense.

> Τὸ κυπαρίσσι τὸ λιανό
> αἰέρας τὸ ζευλώνει,
> τὴν κοπελιάν τὴν ἀγριάν
> ὁ νέος τὴν ἡμερώνει.

[10] Vol. I. pp. 251—253. Compare THEOPHRASTUS, H. P. II. 2. Κυπάριττος δέ, παρὰ μὲν τοῖς ἄλλοις, ἀπὸ σπέρματος, ἐν Κρήτῃ δὲ ἀπὸ στελέχους, οἷον καὶ ἀπὸ τῆς κουρᾶς, ἐν Τάῤῥᾳ· παρὰ τούτοις γάρ ἐστιν ἡ κουριζομένη κυπάριττος. Compare BELON, in his Observations, etc.

[11] BENTLEY, Epist. ad Mill, p. 25.

[12] See PAUSANIAS, II. p. 180.

[13] DIODORUS SICULUS, V. 76. Βριτόμαρτιν δὲ, τὴν προσαγορευομένην Δίκτυνναν, μυθολογοῦσι γενέσθαι ἐν Καινοῖ τῆς Κρήτης. On this story see Heyne's Excursus on the Aeneid.

[14] PLINY, N. H. XXI. 14.

[15] Above, p. 107.

The meal furnished by the hospitality of my Samarióte guide consisted chiefly of the flesh of a wild-goat, killed by him on an expedition from which he had only just returned. I obtained from him three pairs of the animal's horns[16].

The wild-goat or agrími[17] of Crete, is supposed, by Belon and all subsequent writers on Natural History, to be the bouquetin or ibex of the Alps. This, however, does not seem to be the case[18].

The following account of the animal is a close translation of the very words used by an intelligent mountaineer[19] in conversation with me. "The agrímia are so active that they will leap up a perpendicular rock of ten to fourteen feet high: they spring from precipice to precipice, and bound along with such speed that no dog would be able to keep up with them, even on better ground than that where they are found. The sportsman must never be to windward of them, or they will perceive his approach long before he comes within musket-shot. They often carry off a ball, and, unless they fall immediately on being struck, are mostly lost to the sportsman, although they may have received a mortal wound. They are commonly found two, three, or four together: sometimes a herd of eight and even nine is seen. A party of four Therisiótes killed two wild-goats about 1819, one of which weighed twenty-eight

[16] They were all of nearly the same size. The length of each of those engraved at the head of this chapter is, on its outer edge, 2 feet 7½ inches, and, on its inner edge, 2 feet 1½ inches.

[17] See above, p. 232. note 4.

[18] As appears from an examination of the horns. My friend Mr Rothman, Fellow of Trinity College, Cambridge, writes to me, on examining them, "it is not the bouquetin, to which however it bears considerable resemblance, but the real wild-goat, the capra aegagrus. Pallas. the supposed origin of all our domestic varieties. The horns present the anterior trenchant edge, characteristic of this species. The discovery of the aegagrus in Crete is perhaps a fact of some zoological interest, as it is the first well-authenticated European locality of this animal."

[19] Captain Vasíli Khális, of Thériso.

okes and the other thirty-five. They are always larger than the common goat. In the winter-time they may be tracked by the sportsman in the snow. It is common for men to perish in the chase of them. They are of a reddish colour, (*κόκκινα*,) and never black or party-coloured like the goat: the number of prominences on each horn indicates the years of the animal's age."

The great activity of the wild-goat is alluded to by the epithet *ἴξαλος*[20] bestowed on it in Homer; and the length of the horns, which I obtained at Samaría, is very nearly that assigned in the well-known description of the bow of Pandarus[21].

Before four o'clock I set out on my return to Haghía Ruméli. Again and again do I "cast a longing, lingering look behind," as each step which I retrace, towards the narrow glen, makes me lose sight of the lofty cypress-clad mountains of Samaría.

> I mark the sable woods
> That shade sublime yon mountain's nodding brow:
> With what religious awe the solemn scene
> Commands our steps!
> As if the form of Minos should forsake
> Th' Elysian seats, and down the embowering glade
> Move to the pausing eye!

On leaving these grand and most beautiful of Nature's works, it is not without a feeling of regret that I have

[20] On which see HESYCHIUS, v. Ἴξαλος. EUSTATHIUS, on Il. IV. p. 342. VISCONTI, Museo Pio-Clementino, Tom. III. p. 54. HEYNE, on Hom. Il. IV. 105.

[21] HOMER, Il. IV. 105.

> *Ἀυτίκ' ἐσύλα τόξον ἐΰξοον, ἰξάλου αἰγός*
> *ἀγρίου, ὅν ῥά ποτ' αὐτός, ὑπὸ στέρνοιο τυχήσας,*
> *πέτρης ἐκβαίνοντα δεδεγμένος ἐν προδοκῇσι,*
> *βεβλήκει πρὸς στῆθος, ὁ δ' ὕπτιος ἔμπεσε πέτρῃ.*
> *τοῦ κέρα ἐκ κεφαλῆς ἑκκαιδεκάδωρα πεφύκει.*

where HEYNE observes, (Tom. IV. p. 573.) "Grammaticus apud Venetum A. XVI. palmos aestimat *δύο ἥμισυ πηχῶν*. Voluit puto dicere *ποδῶν*. *duobus pedibus cum dimidio.* Scilicet pes Graecorum censetur VI. palmis, *palmus* autem quatuor digitis. (Palmus est *δῶρον, παλαιστή, δοχμή*—) Tam magna esse caprarum agrestium cornua olim dubitatum est."

only been allowed to gaze on them for a few hours, and in all likelihood shall never again behold the glories which are now so rapidly vanishing from my view. I am indeed leaving

> A land whose azure mountain-tops are seats
> For gods in council; whose green vales, retreats
> Fit for the shades of heroes, mingling there
> To breathe Elysian peace in upper air.

END OF VOLUME II.

HISTORICAL APPENDIX.

HISTORICAL APPENDIX.

Obstacles having arisen which prevent me from executing my original plan of writing not only Travels in Crete, but also a History of the island, I think it well to publish extracts from some of the authorities which I found among the Manuscripts of St Mark's Library at Venice.

I will begin with some passages of old Venetian chroniclers, which throw light on the history of Crete during the thirteenth and fourteenth centuries. The Manuscript with which I will commence is of the fifteenth century, written in a hand very difficult to decipher, and is in the Venetian dialect: it consists of 117 leaves, is in the quarto form, and is numbered "Codice XVII." in the catalogue of St Mark's Library. It is called "Cronaca Veneziana, dal principio della città sino al MCCCX."

Fol. 73. "Piero Ziani Doxe per elecçion monta a questo dogal dominjo chorando li anni del nostro Signor Miixiiij."

He arms thirty-one gallies, of which Miss. Renier Premarin was appointed captain, "lo qual dispartandosse de Veniesia chon quela armada, lo chastello de Chorfu e de Modon e de Choron per força darme elli prese, e andando plu auanti elli prexe naue iiij de çenoexi in Spinalonga, e puo navegado alixola de Crede elli prexe lio bon chapello Chorssero chon alter pluxor çenoesi, li qual' iera chó gallie iiij e naue vj. e tornando lo dito Chapetanio in driedo, lo dito Chorsser a Chorfu fo apichado per la golla chon molti suo Chonpani insenbra. Anchora li diti Chapetanj avanti che elli tornasse in driedo *elli per força darme la çitae de Chandia e tute le ÿsole e chastelli, e forteze, con tutti quelli de Crede elli prexe*, e fo mandado in quella fiada, ducha in Chandia Miss. Iachomo Tiepolo el savjo la qual chosa per hordinamento del dito Doxe la ijsola predita de Crede ello devissa e parti entro tuti li nobelli de Veniesia e chi volesse andarla ad habitar dipendando e quella vegniva chamada Chavallarie chomo anch Crede e ave tuto quello che per la Signoria li fo dado alluy e ala so famiija.

Fol. 78. "E comença ladita vera in ann. M.II.LXXXX. e fo Chapetanio di gallie LXVI. Miss. Ruçier Moresini." He recounts some losses of the Venetians at sea. "Anchora durado ladita vera, çenouesi prexe la Chania che lysola de Crede e finalmente de un tratto da le partti si ando mecharçiendo la pace fo fatta entro lor."

Fol. 81. "Anchora in quello tempo quelli de lisola de Crede revella al chomun de Veniessia per la qual cossa el ducha che iera in quel tenpo in Chandia e fo Miss. Zustignian Zustignian aparechia subitamente uno nobele exerçito de zente da pie e da chavallo e dette la bataia per tal che in puochi çorni ello redusse i diti revelli a fedeltae de chomun de Veniesia e questi fo li Chalergi."

Fol. 83. "E in quello tenpo alguni nobil griesi de Crede che non stavva in la çitae fe grandissima assurança de li suo griesi dela detta ysola reuelando al chomun de Veniesia e a tute le çitae e chastelli de ljsolla redugandosse a lo mazor forteçe de lisolla e stagando soura le coste della montagna e domiçando molti luogi della Chavalarie de Chandia e da tute quele partte vexine, la qual rebelaçion siando vegnuda ale orechie dello exçellentissimo Doxe, et per xente chon grande solenitae elletti e mandai ala dita isola pedoni e ballestrieri chon gallie de çipro le qual iera per partir e per andar a marchado ço su li nobelli e valorossi homeni Miss. Nicholo Falier Miss. Zustignian Zustignian Miss. Andrea Morexini el Chavallier li qual fose provededori e testadorj e chapetani de la dita vera. per tal che —— li diti chaporalli chauui de questa rebellaçion fo prexi e dadi in força delli diti Signori li qual unicamente chomo a rebelli e pessimi traditori e produtori elli fe ad alguni tajar le testte façandolli morir crudielmente, e altri alguni si li fe força in mar sichomo se chonvegnivva; li qual altri griesi del puovolo menudo, retorna puuo ala soa fedeltae e abiando messo li provededori, tuta lissola in paxe elli retorna puuo a Veniesia do de essi e fo reçeuudi chon grandissimo honor ello terço che fo Miss. Nichollo fallier paga el debita de la humana natura ela in Chandia el fo sepellido.

Fol. 96. (2). Anchora in M III LXIII del mexe di setenbrio vene novelle a Veniesia chon una galia del golfo ço fo Miss. Nichollo Fallier Sava Chomito digando chomo li çentillomeni de Chandia si aveva tolto çoxo del palaço lo ducha e li chossierj e tuti li altri ofiçiali che per Veniesia se apellavva de la soa Signoria per alguni daçii li qual el chomun de Veniesia aueua mandado.—— e tuti fo metudi in presion. Anchora li çencillomeni de Chandia si fexe elli uno per so ducha e retore de la tera per chomandamento de tuti Miss. Marcho Gradenigo dito Spiritello, e si messe in pression Miss. Lunardo Dandolo che iera ducha chon tuti quelli che se trouvaesse de Veniesia e fexe retegnir le nauue e nauuilli da Veniesia ço i paronj e nocheri e marineri le qual naue se trovaesse charegade alle vendeme. Anchora in an M III LXIII del mese de setenbr. se arma v gallie ço fo iij in Veniesia e una a Clotià e una in Chavo d' Istria e fo fato iij nobeli çentilhomeni provededori per meter la tera in paçifico stado. Anchora in M III LXIII del mese de setenbrjo fo fato v nobelissiml çentillomeni ambassadori: li qual andasse a proveder in Chandia, e rechonçilliarlli chon lo chomun de Veniesia e metterlli in paxe e in achordo— e li diti provididori siando çonti in Chandia e vojando essi a parlare contro de la dita revellaçion apresentandoli lelletere del chomun de Veniesia per li çentillomeni de Chandia no li volse essi dado oldienza alguna elli diti V Savj vene in Veniesia cholle gallie de romania le qual vegnivva da marchado e chonta la novella ala dogal Signoria la qual non li plaoxete niente.

In quello tempo de Miijlxiiii al.... de març o siando fato Chapetanio general de tera el nobelle homo Miss. Luchin dal Vermo ello fo de zente per tuto Veniesia e puo dover asunar una gran quantitae de moneda e questo fo per soldar la dita zente da pe e da chavallo e ballestrieri assai; li qual siando tute asenbladi in Veniesia subitamente in puochi çorni fo armado gallie sotil e arsilli grandi per portare li diti chavalli con li soldadi, e altri navillj e navve assai che li porta, e fato la mostra de li diti soldadi da pe e da chavallo el nobile homo Miss. Luchin dal Vermo chapetanjo fexe chargar li chavalli entro li arsilli e apareciarggi de tuto quello chelli bisognava e inllo nome de Dio e della Verçene Maria e del Vangelista Miss. San Marcho die X Avril M III LXiiij el se parti del portto de Veniesia con quelli arsilli e gallie e legni sotil e nave che aveva armado la dita Signoria, siando Chapettanio de mar Miss. Domenego Michiel de Santa Sofia e fu.... in mar façando far vela e navegando die e notte chomo pllasette 'a Dio elli çonsse ala Frasschia a vij di de magigio ella stette tuto quel di chon la dita armado chonzando le suo chosse a mejo che lj sepe e puuo laltro di che fo a viij di magigio elli desschar li suo chavalli in tera chon tute le suo arme e arnexe valenttemente al despetto delli suo innemixi Chandioti façando pavessa tuto intorno la soa hosta e siando vegnudo l' altro di da mattina ço fo alli viiij magigio el nobello homo chapettanio de terra mess. Luchin dal Verme si fe sonar la tronbetta per tuta la hosta ello istesso monta a chavallo tuto armado chavalchando per tuto loste dagando hordene alle schiere e a quelli da pe e da chavallo e ali ballestrieri metandolli alle suo poste façando pavesade e chontraxiando a tuti li suuo innemiçi e fe so banderaro pasquallin minotto lo qual fo ardido e possente çovene elleal senpre portandosse vallentementte ala dita bataja e abiando dado ordene a tuta loste lo dito chapetanjo Miss. Luchin dal Vermo chomo sapientissimo he maistro de guera ello fexe clamare bandir per tuta la hoste che tuti li chaporalli e chontestabelli de da pe a da chavallo vegnisse alla soa tenda, e chosi fo hobedido lo so chomandamento, e siando vegnudi tuti davanti llui ello li fe sentar tuti apresso de si ella in quello luogo elo fexe uno parlamento molto soavve e bello, digandolli chosi: "Fradelli mie charissimi io ve priego tuti quanti vui li qual se quua al prexentte che vuuj sie llialli e fedelli ala dogal Signoria de Veniesia le qual Sign. ha mandadi quua questo a far e per tanto se vuol che vuj sia prodomeni e valenti a chonbater e no ve infença del ben far che io ve inprometto per lo sagramento, che io ho fato a Dio e al chomun de Veniesia, che quanto per mi vojo esser lo primo ferido   un bon sochorsso de quanto io poxo se Dio me impresta tanto de vita che io no me infençio miga del ben fare che avanti io vojo morir chon honor che viver chon vergogna sapiando che se de Dio ne da la graçia che nuuii abiemo vittoria delli nostri nemici, la Signoria de Veniesia e ben si sauvia e si chortexe che li ve radoplera doplamente del ben e de lo honor che vui li avera fatto, e se vui farete lo chontrarjo che vuij non fosse llialli io ve inprometo che chollui che fallirà sara inponido, io faro la çustixia se el fosse mio fio propio, siche in penso de ben far, altro io no ve jo diro mo al presente perche ello tempo de fare fati." e ditte parole ello fo fin a lo so parlare e tuti se refresscha e mança   e abiando tuti mança e bev. Miss. lo Capettanio si fe sonar la tronbetta e tuti monta a chavallo ben armadi e chomença a chavalchar in so la çitae de Chandia chon grande ardimento molto ben asscheradi e quelli della tera ensi fuora inchontralli nostri molto ben a ponto da pe e da chavallo e la schomença la bataja mollto crudiel e pericoloxa essende morti e feridi plluxor deli grexi per tal che elli non potte pllus durar ala bataja e si delle spalle allini ritornando in verso la çitae; ello nostro Chapetanio Miss. Luchin dal Vermo pasqualin Minotto che jera banderaro chontinuadamentte li seguiva tajandolli e olçidandolli, e puochi fo delli griexi che tornasse vuivvi in la çitae che non fosse morti davanti la porta da Chandia. E chomo piaxette a Dio Veniçiani   sovra li Grexi in quello di e questo fo in an MCCClxiiij a die Viiij magigio e puo a die x del dito mese elli ave la tera liberamente ço fo Chandia, intrando dentro tajando e olçicando molltij delli traditori chelli trouva e montando in sul palaço elli olçixe Miss. Marcho Gradenigo de Chandia che iera fato so ducha e tolsse via la sua insegne de San Titto e messe quella del Vangellista San Marcho chon gran fessta e alegreça."

The great engagement between the Venetians, under Lucchino dal Vermo, and the revolted Cretans must have been near

Palaeókastron, and perhaps partly on the hills near Armyró, seen in the view at p. 258, of Vol. I. The contest is also described in Codice XIX of St Mark's Library, p. 52.

"A di 10 dito, con il nome de Dio el dito Capetanio schomenzo a cavalchar verso la cita de Candia et schomenzo una dura et aspra batagia et li nostri ferendo et ocidendo li Grecj li ando incalzando in fino ale porte dela tera, et qui fo una grandissima ocision da una et altra banda ma ale fin, como piasete a Dio et a M. San Marco, li nostri intro in la tera, per forza, et amazando quanti li vegnivano incontra: et tanto loro fecero che esi ando ala piaza et intradi in palazo loro trova M. Marco Gradenigo che li rebeli havevano fato dose; et li nostri el tagio a pezi et da poi schorsa tuta la tera, per nome de San Marco. Da poi tornadi in piaza el fo butado a tera la insegna de San Tito et fo levado quela de San Marco."

From Codice xx (di fogli 233, del secolo xv. in circa; vedi Il Catalogo della Biblioteca Marciana, p. 227.)

Fol. 62. "Haven° compra la Signoria de Venetia la Isola de Candia dal Marchexe de Monfera mando a tuor el dominio de la dita isola onde per greci fo vidado e contradito di volerse dar segondo i pati del suo Signor. E vedan° questo Miss. Renier Premarin, cap° di mar de larma de la Signoria, se misse in ponto, e conbatti valentemente la dita cita de Candia, e per forza quela prese, e per lo simel tute le so altre cita, e castelli, primte e jurisdiction, sotometandoli ala Signoria di Venetia fazom° quelli luoghi di gran robarie, e metando tuto a sacomano. Dapoi dita armada molto richo torno a Venetia, e questo fo del 1211, e fo deliba di partir la dita isola tra Zentilhomeni e populari, che volesse andar ad habitar in quella isola con le so femene."

Fol. 65. "Del. 1214. In tempo di Miss. Piero Ziani li usl de Zenoa uno de bassa condition chiamado Rigo, homo valentissimo e savio, e tanto fexe con soa industria che l' have gran parte del dominio de la Isola de Candia per la qual caso el fo chiama el Conte Rigo de Malta. In questo tempo era re de Zezilia re Ferigo che havea pocho poder ma de poi vene in gran posanza. El dito Conte Rigo prese con una so nave una de' Pisani, e dapoi con quelle do nave una di Venetiani, e poi torna a Malta e missesse benissimo in ponto metando su queste tre navi gran quantita de zente e de victuari, e da poi prese una nave grossa richissima de Venetia che vegniva da Constantinopoli: et havendo aguadagnato grande haver, have modo di haver gran quantità de zente sorti a danizar, e ando alissola de Candia, et appandose li greci lo accepto in la dita isola, dandoli gra pre Signoria. E subito el dito Conte Rigo feze far de molti castelli per potersi prevaler da la potentia de la Signoria de Venetia. El primo castel fo el castel de Mirabelo, uno de Monte-ferat, e uno che se chiama Bonifacio, e Castel Nuovo, e Bel Riparo, e molti altri castelli[1], siche i fo per n° xiiij° castelli; anchora tegniva el Castel Termene e possedeva la cita de Candia e la Chania et have animo de mandar a Roma a dir che j lo incoronasse re de quelle isola, e sel se non si havesse trovà un capelan de Venetia che saveva ch' el dito Conte Rigo jera anda furtivamente a tuor deta isola, e desse notitia al papa l' havaria habu . . . . . . Donde sapuda che have el papa la cossa, cazo j suoi ambassador via con bruta vergogna. E sapudo che ave M. Piero Ziani doxe delli Venetiani che Conte Rigo havea occupa lisola de Candia, e prese per parte de armar subito xxx^a galie e otto nave de le mazor de Venetia, e fo Cap° de la dita armada Mess. Jac° Baseio e prontamente el se parti de Venetia, e fato Captanio de le galie de marchado Miss. Renier Dandolo, e Miss. Renier Premarin, e tuti andono de compagnia, e questa armada prese Lio Vechio Corsaro, nemigo de Venetiani, e prese Rigo da Bela pola Zenoexe, con galie xij e cinque nave, perchè Venetiani e Zenoexi erano capital nemixi; e da poi lo dito Cap° con le 30 galie e 8 nave e altri navilij assai, con le galie da marchado, se ne ando alisola de Candia, perche el dito Conte Rigo haveva fornido xv castelli de molta zente darmi per esser Zenovese, Zenovexi li avea da granda aidia per esser in guerra con Venetiani, e fo dade molte bataie e niente se pote far, istando alassedio della dita isola el fo preso ala batajo Miss. Renier Dandolo Cap° de le galie de marchado, e fo mena al conte Rigo e fo messo in preson, e siando disdegna el dito Miss. Renier, et non pote mai manzar e mori in preson,

[1] See Vol. I. p. 64. note 8. Vol. II. p. 40. and p. 73.

e da poi fo fato de gran bataie. Miss. . . . Premarin Cap° vene a Venetia per esser compido el so tempo deputado havendo fato valentemente, e in so luogo fo fato Miss. Polo Ziani, el dito Cap° fece guerra valentemente al dito Conte Rigo e cominzo a tuorli di suoi castelli, e vedando il Conte Rigo esser astretto ocultamente lando a Zenova in persona a domadar soccorso, e quelli da Zenova li de gran soccorso di cavalli e pedoni e delli quattro grossissime nave e sie galie Cap° M. Jac Longo, e siando vegnu dita arma per demontar ala isola di Candia, larmada de Venetia li de driedo donde se messe a scampar, e vegando al note adoso con segnali de fuogo ingano larma de Veñij che non sape andar dredo, e scampo da quella el Conte Rigo con una galia vene alisola de Candia prestamente scapando larma di Venetiani, e dismonto in terra, e la matina larma de Venitiani vedando aver pso quella de Zenoexi i se ne torno in Candia, et breviter el Conte Rigo vene a pati de doverse render a Venetiani, e fra li altri pati fo che dovesse dar a Venetiani tuti li castelli ch' haveva su lisola di Candia e renuntiar tute le ragion chavesse in dita isola e —— chel fosse tegnudo a mandar un so nevodo a Venetia, e che la Signoria desse per dote al dito so nevo perperi 1500 e desseli una Zentildonna, e a questo modo la Signoria have lisola."

Fol. 72. "Del 1299 Zenoexi prese la cita de la Cania che e suso la isola e quella bruso a robo, e al fin siando strachi una parte e laltra Zenoexi e Venetiani fecero pase."

THE REPORTS of FOSCARINI and GARZONI to the Venetian Senate, on the condition of Crete towards the latter end of the 16th century, have been frequently referred to by me in both this and the previous volume. Foscarini had been preceded by BASADONNA, a small fragment of whose Report is still found in the Library of St Mark. It is numbered Classe VII. Codice CCCIV., and is entitled "Relatione del E^{mo} Sr P^{o} Basadonna, del Sindicato de Levante, nel 1566." He says, at fol. 2, that such visits of Syndics "Furono con tanto frutto et beneficio publico et con tanta consolatione a quei miserabili paesi, che, non essendosi già stati, molti et molti anni, Sindici in quelle parti, restorono *sollevati da molte grandi et crudelissime oppressioni et tirannie.*"

Again, "Un certo dazio che chiamano dei Parici, appare a molti poveri e più miserabili di quel territorio infima calamitá —sono astretti fuggire in Turchia, perciocchè essendo costoro obbligati pagare in camera un aspro per testa all' anno, et in caso di morte lasciare tutti i beni loro al .... morendo senza figliuoli maschi, i Daziari, et altre genti da loro dipendenti, usano contro questi infelici, et nell' un caso et nell' altro, strane et horrende tirannie—Sarebbe cosa degna della sua clemenza, subito levare una così odiosa et barbara esazione—poichè il conservarla è sottoporre quei miserabili huomeni a crudelissimi serpenti, che lacerino et li divorino afatto, o li sforzino, li pochi che sono avvanzati, *fuggir subito in Turchia, seguendo l' orme d' altri innumerabili, che per questa cagione di tempo in tempo ne sono andati.*"

I will now give some extracts from GARZONI's Report, which is numbered "Classe VII. Codice CCCIV." in St Mark's Library, and is entitled "Relatione del Ecc^{mo} Sr Giulio de' Garzoni del Sindicato de Levante, nel 1586."

Fol. 9. "Possede questa Eccelsa Repubblica Seren. Principe Illmo et Graviss^{o} Consiglio, l' Isola di Candia, degna per il numero degli habitanti, per la nobiltà della Colonia, et per la qualità delle fortezze, che cingono quelle città per sua difesa, d' esser chiamata Regno."

"Parlerò dei popoli et bisogni loro, et loro voluntà, delle spese, fatte dalla Serenità Vostra, per provisione di guerra, et l' entrate rese dall' Isola, et del modo che tengono li Magistrati nel governo del Regno, dalla giustizia de' quali dipende il pronostico di possedere felicemente quello stato.

"Nelle quattro Provincie sue, che sono di Scithia, Candia, Rettimo, et Canea, fuori dalla città si trovarono anime centosettantasei mille quattrocento trentatre[1], delle quali sono da fazione cinquanta quattro mille settecento ottantasette.

[1] The population had been long decreasing. FOSCARINI, Relatione etc. fol. 133, gives 1070 as the number of villages (casali) and 219,000 as the number of souls, of whom 55645 were "da fattion," 29218 of them residing in the cities, and the remaining 26427 in the country. The codex No. 766 in the Racolta Correr at Venice gives the following table, the amount of the population somewhat earlier.

Megalo-

"Di questro membro di cittadini, principaliss° del Regno, darò conto assai particolare, et credo esser bene che le VV. SS. Ecc^me^ comportino, con qualche patientia, il mio ragionamento, potendo essere, che le siano per intendere cose, e nuove, et a proposito per edificare quelle genti a devota fede verso la Serenità V.

"Altre volte quelli popoli furono possessi da particolari Signori, venduti[2], comprati, et comandati come Parici[3], et la Colonia che mandò la Serenità V. successe al dominio de' beni et habitanti suoi, la quale tenne il medesimo stile per lungo tempo: in maniera che le genti, fatte servili per lungo habito di servitù, hanno convenuto riuscire, e vili et senza industria.

Fol. 10. "Hebbero questi Governatori a principio qualche difficoltà a far lasciare l' arco a quelle genti[4], arma sua antica et naturale, per farle prender l' archobuso: ma, havendo alcuno

	Villages.	Priests.	Monks.	Men and Children.	Women.	Total.
Megálo-Kástron	496	492	570	48628	49984	100170
Setía	148	256	268	9255	9254	19181
Rhíthymnos, City		38		3862	4138	8038
Villages	296	345	410	34694	25590	61335
Khaniá, City		52		5130	5271	10453
Villages	314	412	516	31216	28560	61018
Akrotéri		36		1819	1533	3388
Sfakiá		28		4228	3650	7906
	1254	1659	1764	138832	127980	271489

[2] The purchase and sale of *slaves* prevailed at Christian Venice in the fourteenth century. See GAMBA, Serie degli Scritti impressi in dialetto Veneziano, Venezia 1832, who gives at p. 35. an "Istrumento di vendita d' uno Schiavo, scritto l' anno 1365," and observes, "Notabile riescirà il leggere un vecchio Documento di *jus servile* presso i Veneziani, da cui apparisce espresso il consenso di uno Schiavo di passare da un padrone ad altro."

[3] FOSCARINI observes, Relatione etc. fol. 107. that although, when he wrote, that generally notorious and odious "parichia" which existed under the Venetians in Cyprus, did not prevail in Crete, still even here there existed some "*parichi*" by descent, who had still this "obbligo et servitù di parichia," and that all the peasants were obliged to perform "tante angarie," and that some "obblighi" had been introduced of such a kind, "che non saprei se fossero del tutto liberi."

[4] See above, p. 254.

dato principio a valersene alle caccie, è stato di qualche allettamento agli altri, sicchè si lasciarono intendere il desiderarne, et essendo compiaciuti dalla Serenità V. ne furono mandati certo numero, li quali si abbaterono a capitare mentre noi eravamo alla visita sua, et furono dispensati per le nostre mane, accompagnandoli con parole di confidenza, per incorargli alla difesa pubblica, et per introdurgli affetto et hobbedienza alle cose comandate dalla Serenità V. Gli accettarono con allegro animo, et si deve per ragione aspettare ottima riuscita da loro con quell' arma, essendo essi per complessione calda et secca, che gli fa riuscire presti, gagliardi, et vivaci, sono avezzi al patimento per la strettezza del paese, pieno di balze, tutte condizioni accomodate all' archibusiero : sicchè si può credere con fondamento che, per le qualità di quei corpi et per quelle di quel paese ove hanno a maneggiarsi, poco numero di quei soldati, ben disciplinati, sia per travagliare gran quantità di forastieri, potendo loro aggiungere, et fuggire a suo piacere[5], et con poco bisogno di vittuaria, vivendo loro d' aqua et herbe con poco pane. Hanno per suo costume il portare certo sacco alle spalle, dove portano i suoi bisogni in maniera accomodato che non sentono alcun impaccio."

Fol. 12. "Fra queste milizie descritte vi sono li habitanti alla Sfachia, nazione più ardita del resto dell' isola per maggior ferocità di complessione, et per natura di sito, et perchè ancora credono, per certe loro memorie che dicono havere, esser discesi da' Romani, et perciò, con quella credenza, hanno non so che di generoso nelle sue attioni, ne sopportarono al principio il giogo de' Cavallieri, come il resto de' cittadini, o per sua risoluzione, o vero per la sterilità del paese suo non invitò la Colonia a mettervi il piede: ma, sia come si voglia, non hanno alcuna soggettione a quell' ordine di Signori, tanto abborrito dal Regno.

"Si sono quelle genti alle volte mostrate disobedienti alla. Serenità V. et hanno convenuto li Magistrati darli sacco, et tagliarne a pezzi molti, et altri levare dalle proprie habita-

[5] This is just what the Sfakians did in most of their contests with the Mohammedans between 1821 and 1830.

zioni, come fece l' Ecc^{mo}. Cavalli[6], con giusta causa nota alla Serenità V. E certo, se non fosse che tutta la Sfachia non è habitata da quelle fameglie pretendenti la insolentia del disobedire, et che tra essi Sfachiotti non sono uniti, difficilissimo sarebbe haverli in obbedientia: ma questi duoi contrarj, et l'esempio delle cose passate, gli indebolisce in maniera che hanno alquanto mortificato l' orgoglio. Sono partiti in due prole l' una detta Patera, e l' altro Papadopula, discese, per quanto dicono, da duoi fratelli, l' uno Patera, che vuol dire capo di monaci, e l' altro Papà cioè prete, et si generò per concorrentia inimicitia tale, che, aumentandosi sempre, sono fatti gli animi suoi implacabili.

"L' Eccell^{mo}. Foscarini li visitò et li instituì alcuni ordini per conservarli in obbedientia del Ch^{mo} Rettor della Canea, li gratificò di alcune concessioni, et gli deputò ancora Cap^{o}, per la sua esercitatione, facendoli descrivere nei libri delle battaglie come privileggiati; in maniera che il spavento delle cose passate, la induglienza dell' Ecc^{mo} Foscarini, et la forza delli ordeni posteli gli hanno ridotti a qualche obbedientia. Ma la causa principale, per quello che ho potuto scoprire, della sua alienatione, precipitando nella desperatione, è tutta indústria de' Cavallieri, per l' odio che portano a quella gente, come quella dalla quale non possono trahere alcuno utile, et perciò usano ogni potere per rappresentarli alli Ecc^{mi} Rettori per licentiosi, ladri, et homicidiali."

Fol. 13. "Non contenti li Caval^{ri} di aver usata l' industria, con li Ch^{mi}. Rappresentanti, per farli concepire mali pensieri contra li Sfachiotti, sono altrettanto industriosi in ispaventare con suoi ragionamenti quegli huomeni per rigore dell' ordine, et così li fanno disobedire; per il qual mancamento li sono poi fulminate adosso proclame et sententie, che gli levano affatto la speranza di esser gratiati. Pare quasi troppo artificio usato da quei gentiluomeni della Canea; nondimeno è pur vero, ma è anco ragionevol cosa il crederlo, poichè dipende da interessi di quel paese interessantissimo, et in questa maniera riescono alle volte gli Sfachiotti suspetti di fede.

[6] See above, p. 154.

"Al nostro arrivo nel territorio della Canea erano passati alquanti anni che non havevano prestata la debita obedientia di appresentarsi, per la diffidenza che io ho detta di sopra. Io di manco anni et di miglior forza di corpo del Ecc^{mo} mio Collega, volsi fare quella visita, salii quei monti, o per dir meglio li penetrai, poichè vi sono alcune strade piutosto voragini, fatte per discesa d' acque, che per uso di viandanti[7], havendo prima, per via d' un proveditore che si tiene a quel governo dal Consiglio della Canea, fattogli intendere la mia andata, per consolarli et gratificarli dove avessi potuto farlo honestamente: et capitato a' confini de' suoi casali, fui incontrato da cinque cento huomeni, tutti armati di archobusi et archi, tra quali vi era la famiglia de' Pateri, al numero di dugento huomeni, li Papadopuli suoi contrarij non si videro per la inimicitia. Si appresentarono quelle genti con riverenza, et io li accolsi in nome della Serenità V. con quelle parole che mi somministrò la mia buona volontà per bene edificarli all' obbedienza de' Magistrati. Et capitato al suo principal casale che è grande, situato, nella discesa di un alto monte verso la spiaggia, dalla banda di Ostro, dove è anco il Castello della Sfachia, posto sopra un sasso al mare, sicuro per la battaglia da mano, nel quale risiede il Proveditor loro, feco chiamare a me li Capi senza la moltitudine, a' quali parlai con qualche austerità, riprendendoli della poca obbedientia prestata alli Ecc^{mi} Rettori della Canea, in esecutione dell' ordine Foscarini. Con molta modestia mi risposero, anzi con tanta, che non solo non l' aspettai da huomeni fieri, rozzi, et sospetti di mala volontà, come furono rappresentati anco a me, ma da huomeni civili, riconoscenti l' autorità del Superiore, non si havrebbe potuto aver più discretta risposta, allegando per sua difesa le cose discorse di sopra. Gli feci capaci dell' intenzione della Serenità V., et di che maniera deve essere eseguito dal Ecc^{mo} Rettore, quando siano obbedienti all' ordine Foscarini, in maniera che accettorono la riprensione, et restorono consolati, et loro stessi mi ricercarono nuovo ordine per castigo di quelli che non saranno pronti a prendere li ladri e consegnarli alla

[7] FOSCARINI observes, in recounting his visit to Sfakiá, where he remained six days, fol. 106. "Il sito et asprezza di quelli luoghi sono veramente accommodati et proprii alla ferocità di quella gente, ch' è bellissima, bellicosa, et brava."

giustizia. Li tenni ragione, et accomodai alcune sue differentie, vidi il modo del governo di quel Proveditore, il qual trovai senza alcuna forma, si per poca obbedienza, come per venalità di quelli che sogliono andare a quei carichi. Mi fecero la mostra, et riuscì bellissima.

"Sono quegli huomeni più alti, et più formati di vita, con faccia di più spirito, del resto de' Contadini, et per la pretensione della nobiltà passata del sangue suo, et per le proprie forze presenti, riescono più ubidienti, et perciò più pronti all' arme, sono più esercitati nell' archobuso per le caccie, delle quali si può dire che vivono.

"Io mi posi fra loro, et li feci maneggiare secondo la volontà del Collonello Emiliani che era meco, mostrando loro piacere di essere comandati da me, et io mostrai di sentire contento di così allegra obbedienza. Li feci tirare a . . . d' arco et d' archobuso dispensando li pretii di mia mano, per meglio manifestare il piacere da me sentito per la sua pronta obbedienza.

"Andai al Castello S. Nichita habitatione de' Papadopuli, e seco feci il medesimo per consolarli, havendo questi ancora causa di maggior disperatione, come più poveri, et inferiori di numero alli suoi inimici.

"Vivono quelle Prole principalmente dell' industria degli animali di pascolo, de' quali ne hanno quantità per l' abbondanza de' monti non usati d' altri vicini, et per la difficoltà della sua ascesa, et per lo rispetto che si fanno portare da' vicini. Gli altri habitanti alla Sfachia vivono di far barche, et navilj, per la comodità de' legnami et pegole che si trovano in questa parte dell' isola; et a tempo di bisogno sono passati al servitio dell' arsenale della Canea.

"Il quarto giorno del mio arrivo parti, lassando ordine alli capi dell' una et l' altra Prole che venissero alla Canea, che gli avrei fatto perdonare il fallo passato con quella obbedienza. Vennero pochi giorni dopo—et all' hora conobbe il Ch^{mo} Rettore le false informazioni dateli da' Cavalieri; poichè a quel tempo non comparse alcuna querella contra gli Sfachiotti, cosa che il fece maggiormente affezionare a loro, et puote confermare li miei officii: et presi gli ordini fatti col consiglio del Ecc^{mo} mio Collega, per dar forma megliore al governo del suo Proveditore, se ne tornarono lieti alle loro case; et essendo poco dopo capitato il Ch^{mo} Cap^{no} Barbarigo alla Canea, desiderando S.S. vedere le mostre generali et però comandate le cernede, gli Sfachiotti che non solevano comparire, o con poco numero, vennero intorno a seicento, et furono veduti con gran gusto di quel Signore."

Fol. 15. (2). "Hora dico in tanti casali, nei quali si habbiamo trovato, habbiamo vedute le case de' cittadini, nella maggior parte delle quali non si è veduto cosa alcuna per l' uso del vestire, ne del posare: et per il vito, sono senza pane o grano, non hanno vino, le donne spogliate, li figliuoli nudi, gli huomeni poco coperti, et le case vuote del tutto, senza alcun segno di habitazione humana. Et quella meschinità di huomeni è obbligata per antico uso al Cavaliere duoi angarie personali l' anno, ch' è il lavoriero di dodici giorni, et convengono ancora lavorare pur, per antica istituzione, il resto di quello che bisogna al medesimo Cavaliere nel suo casale, a otto soldini il giorno, che importa una gazzetta[8], et un quintodecimo di essa perciò introdotto gia ducento anni et non più accresciuto. Sono in obbligo di aver polli, et galline per porta, così li Padroni havendo introdotto questo nome di porta per casa, le qual case sono anco fabbricate dagli stessi contadini, ne alla sua partita hanno altra ragione che portare seco la porta, et perciò li Cavalieri, industriosi del suo utile, hanno fatto ad ogni istantiola una porta, per moltiplicare le regalie. I buoi da lavoro, chiamati donnegali, sono in obbligo di lavorare certa quantità di terreno, per lo quale, seminato o no, bisogna che il contadino paghi la terzaria. Sono ancora i donnegali tenuti a farli due angarie l' anno. Muli, et altri anemali da basto convengono fare due agozi, cioè due viaggi, alla città per il patrone, gli anemali da pascolo pagano la decima, chiamata da loro Decata, et altre mille invenzioni per assorbire tutti gli frutti della terra. Se il contadino ha vigne cioè la superficie (essendo il terreno sempre del Cavaliere) piantate, et allevate da lui, benchè in terreno prima silvestre, conviene pagare al patrone, prima che faccia la divisione per regalia, tornando per antica istituzione il terzo al Cav^{re} et due terzi al contadino, cinque . . . da loro chiamati mistacchi, per vigna, sotto pretesto

[8] Two Venetian *Soldi*.

di quello che mangia il Contadino avanti le vendemie per la spesa del Pattichier, chiamando così loro un certo luogo fabbricato per franger l' uva, et sotto altre inhonestissime invenzioni: et per avvantaggiarsi ancora questa regalia, dividono in tante parte le vigne, che poche rendono più di quindici mistacchi; in maniera che con la fraude fondata sopra la forza si appropriano li due terzi a loro, e lassono l' uno al contadino. Sicchè posso affermare con verità, come affermo per la dovuta riverenza che porto alla Serenità Vostra, non aver veduto in casa di alcun cittadino vaso per custodir vino, segno manifesto di non raccogلierne, et se ne vogliono avere, lo prendono a poco a poco dal Cavaliere, promettendo per ricompenza, oglio e seta, ovvero del medesimo vino venturo, a quelli prezi che detta l'appetito suo: et le medesime invenzioni che servono a gravare le persone et gli animali da opera, et a rapire quelli da pascolo, l' oglio, il vino, et il grano, la medesima industria usan per levargli ancora li altri frutti della terra, benchè silvestri, et di piccola importanza.

"Sono deputati per giurisdicenti di questi territorij, come ho detto, Castellani, Scrivani che servono per Cancellieri et capitanj contra Fures, li quali sono tutti volti rapacemente a' danni di quella povera gente, pegnorando quel poco che alcuno di loro sia trovato havere ascoso da' Cav^{ri}, sotto pretesto di disobedienza della quale il contadino abbonda per la disperazione, poichè in ogni maniera è meschino. Gli Castellani non possono per legge giudicare il valsente di più di zecchini due, nondimeno qualche Ch^{mo} Regimento gli ha concesso autorità fino alla summa di perperi ducento, che sono intorno a zecchini quattordeci: et perchè hanno de caratte otto percento, fanno riuscire ogni causa di ducento perperi per piccola che sia per aver li sedici di caratti, con mille altre invenzioni di estorsioni per mangiare a' poveri. Li capitanj contra Fures, il nome de' quali denota il suo carico, hanno le sue utilità da' ladronezzi, et sempre inventano per trarne utile, caluniando li buoni et escusando li rei, con rovina universale. E però vero che li contadini invilliti per la loro inopia, e portati dalla necessità, commettono qualche furto, ma di poca importanza. Li giusdicenti furono introdotti a tempo della Colonia, e dovevano essere Nobeli Veneti o Cretensi; nondimeno a poco a poco si sono domesticati in maniera gli eletori che hora sono esercitati per sostituti, et sono li Scrivani delli cav^{ri} li quali per lo più hanno comperato la Castellania, e tutti quelli che hanno avuto carico di riformare, hanno regolato prohibendo la sostituzione, et comandato alli Regimenti il fare la Rettoria giusta la forma delle leggi: con tutto ciò non si è potuto ottenere nè l' uno nè l' altro.

" Alcune di quelle Castellanie sono state concesse in vita dalla benignità della Serenità V. et così, come è liberal gratia a quelli che l' hanno ricevuta, così è grave disgrazia alli sudditi di quel Castello, poichè per la lunghezza del tempo, sendo gli altri di due anni solamente, si fanno patroni, non solo usando le estorsioni solite de' Castellani, ma quelle de' Cavalieri ancora; et se non havessimo havuto rispetto alla grazia che ella ha fatto ad alcuno di quelli, per la nobiltà della famiglia, et per le disgrazie patite dalla sua patria, molto inclinando la Serenità V. al suo sollievo, gli haveressimo dato qualche castigo, per il poco rispetto che ha havuto alli sudditi della Serenità V."

Fol. 18. " Quei Capitani per fama universale sono pieni di sete per succiare il sangue a' poveri: nondimeno hanno tanti modi di occultare le sue tristizie, che per esquisita diligentia usata dal Ecc^{mo} mio collega nell' inquisitione non se vi è scoperto altro che uno."

Fol. 18. " Gli huomeni deputati alla Galera sono in continuo terrore d' andarvi et quelli che hanno il modo, benche strettamente, per qualche vignetta, o terreni, o anemali, il tutto gettano spensieratamente a vil pretio per aver denari per pagare l' andiscaro che importerà quindici o vinti zecchini, spesa insopportabile alle loro forze. Li più meschini, impotenti ad aver l' andiscaro, fuggono alla montagna, et poi, assicurati dalli Cavalieri, tornano alli casali loro, tanto più soggetti, quanto sono temidi della giustizia, et con il suo esempio tanto più obbedienti gli riescono gli altri casaliotti, attribuendo l' autorità del salvargli dalla galera agli Cavalieri: et per ciò con questi et altri pretesti di sopra discorsi, se ben per decreto publico è levato il nome della Parichia, convengono vivere nella medessina soggezione.

" E adunque la contadinanza aggravata per le Galere, per le angarie, per le rapacità de' giusdicenti deputati per suo governo et per la crudeltà de molti Cavalieri, aggiuntovi ancora l' estorsioni ch' ella riceve per mille accidenti estraordinarij, essecutioni di debiti civili, visite di Rettori et altri ministri, a' quali convengono dare il vivere a pretii vilissimi."

Fol. 19. "Si che il contadino, di questa maniera vessato, et per tante parti stracciato, impreso da' ragionamenti de' papati è fatto nemico del nome Veneto, nè è capace della distinzione dagli Nob. Veneti loro Cavalieri a quelli che manda la Serenità V. di la, per suo governo, et è ridotto a tale per le cose dette che credo poter dire con verità, escludendo però li privilegiati, che desiderano mutazione di governo: et sanno non poter capitare in mano d' altri che del Turco; nondimeno, credendo non poter peggiorare di stato, inclinano anco a quel tiranico Dominio."

"Ma non mi spaventano tanti mali detti, et altri lasciati per strettezza di tempo, et per altri rispetti ancora, poichè quella massa d' huomeni sono di maniera bramosi del sollievo, con sete cosi ardente di gratia, che conseguita con ogni picciol segno fatto dalla Serenità V. a sua protetione, portando il loro stato di miseria a qualche miglioramento, sono per riuscire li più fedeli figliuoli che habbia questa Eccelsa Repubblica, addolcendo l' amaro delle sue disgrazie con la benignità della Serenità V. Con questo pensiero habbiamo dato qualche principio noi, a farli capire il proposito ch' Ella tiene di sollevarli, castigando i Cavalieri, Castellani, Capitanj contra fures, et altri ministri, condennando fino a cento ducati, et privando di essercitare ufficj pubblici, ch' è il termine della nostra autorità per raffrenarli dal tiranneggiare."

Fol. 21. "Vivono adunque li papati in quella Isola senza capo del suo rito, et con poco rispetto all' arcivescovo Latino; biasmano tutte le sue ationi, predicandole per ingiuria a' popoli con impressione d' odio verso i Franchi, et però dalla malvagità et ignorantia di quegli Ecclesiastici non si può dissegnare alcuno conciliamento, senza darli capo a loro superiore, del medesimo ritto Greco, et a me piacerebbe che si gli desse nome di Vescovo, la qual cosa riuscirebbe di somma consolatione alla moltitudine, et se il Vescovato si potesse avere in persona soggetta, et dipendente dalla Serenità V. crederei &c."

Fol. 21. "Sarà ella mirabilissimo istrumento a conciliare l' animo di quelli che per differentia di rito, sono infetti di mala volontà, et a cacciare le superstitiose disseminazioni de' Papati, castigando ancora le persone senza scandalezzare le genti a loro soggette, come occorre ad ogni minimo atto di riprensione, nonchè di castigo, dell' Arcivescovo Latino, che pur ha voluto ponere qualche freno, benchè giusto, a' Clerici di quel rito: nè si può sospettare altro cattivo effetto dal governo del Vescovo Greco alorchè non potrebbe la persona costituita in quella dignità, con tre et più mille ducati d' entrata, migliorare di conditione per qual si voglia revolutione: et havemo esempio, sebene di cosa piccola, assai chiaro, nelle città ove sono i protipapati eletti dal Regimento, perciochè li Preti di qualche considerazione sono tutti ad adulare, nonchè compiacere li Rappresentanti della Serenità V. per ottener gratia di quel loco quando occorre la vacantia, et però nelle città non si ha alcun disgusto de' popoli minuti per cattivi officj de' papati come occorre ne' territorj."

Fol. 26. (2). "In Candia li Cretensi sono molto humili, sendo stati sopramontati dalla forza de' Veneti, et quasi che si possono nominare nell' ordine del popolo. Hebbero altre volte forza e riputatione per pretensione di antichità, et facoltà che possedevano, ma ora fatti poveri, dicono loro per la potenza de' nobeli Veneti—dicono essere arrichiti quelli per possedere la maggior parte de' feudi in feudi, et per gli augumenti de' precii per le militie, dove hanno possuto dare consumo alle sue entrate senza consegnarle ad altri per mercantarle in altri paesi, come si soleva fare prima, et era quella l' industria de' Cretensi. Per le richezze adunque, congiunte con la nobiltà. si mantengono avantaggiati, et rispettati, et dalli Magistrati ancora favoriti, in maniera che non possono più aspirare li Cretensi di tornare alla conditione della Colonia, se non di equalità almeno forti a' loro vicini di riputazione. A Rettimo sono più accomodati di facoltà li Cretensi, et in maggior numero delli Veneti, et perciò più stimati, partecipando loro ancora con qualche vantaggio degli officj della Città. Alla Canea posso dire di esservi equalità, tra l' uno et l' altro ordine, di rispetto, sebbene vi sono maggiori ricchezze ne' nobeli Veneti; di numero vi è poca differenza, ma le discordie di quella città nate per superbia la tiene in continuo travaglio."

Fol. 47. "L' Isola abbonda di terreni, se ben montuosi attissimi al grano, et fanno rendita di dieci, dodici, et quindici per uno. Ho veduto campagne grandi et piane ancora atte a fertilità, se fossero coltivate, non solo per li habitanti loro, a' quali manca il vivere per soli tre ovver quattro mesi all' anno, ma ancora per mandarne fuori. Il ridurla a coltura è difficile per doi contrarij, l' uno che ognuno è volto all' utile delle vigne et ulive, che fanno riuscire l' Isola abbondantissima di vini et oglio, non solo per l' uso suo, ma per la navigazione, come è noto ad ogniuno[9], l' altra l' infingardagine de' villani, nata dalla poca speranza d' adunare cosa che gli resti, come ho discorso. Le vigne[10] et olive per l' utile che rendono a' Cavalieri li distraggono dal pensiero del grano, per la viltà del pretio suo, nè si dobbiamo maravigliare, che ogniuno applichi il suo terreno al maggior utile, per meglio accomodare la famiglia sua, poichè è commune questo desiderio. Fu per l' Ecc^{mo} Foscarini[11], con

[9] This exportation, (see above pp. 52—55.) spread the fame of the Cretan wine far and wide: Iodocus a Meggen, Peregrinatio Hierosolymitana, p. 57. "Nam quid de vino dicere attinet, *cum vina Cretica toto orbe celebrentur?*"

[10] There was a considerable quantity of raisins as well as of wine annually exported from Crete till the outbreaking of the Greek revolution. The following sentences are found in a letter of Monsieur Bertrand, written at Khaniá, and addressed to the English Ambassador at Constantinople, dated 19th September, 1801. (It is preserved among the records of the Austrian Vice-Consulate at Khaniá.) "On fait très peu de raisins secs dans le district de la Canée, mais Candie, et ses environs, en fournissent, année commune, au delà de cent mille quintaux. La récolte des vins qui se fait à Candie, peut s'évaluer, année commune, aux environs de 200 mille mistaches de huit ocques l'une. Les qualités y varient: il y en a de doux, et de l'âpre, mais en général assez bon. Celle des vins du district de la Canée en donne beaucoup plus: on en évalue la quantité à 300 mille mistaches de 9 ocques, année commune. Ces vins ont tous un filet de douceur. Ils sont rouges, le prix ordinaire est de 7 jusqu'à 10 parats l'ocque, suivant la qualité. Il en est du même du prix des vins de Candie."

[11] Foscarini, like some other advocates of corn-laws, seems to have thought it important that a country should produce its own bread, at any price, rather than buy it, at however cheap a rate, if grown elsewhere: he says at fol. 108, of his Relatione etc. that it had been his object "tentar ogni cosa et far ogni esperientia per che Candia nutrisca li proprii suoi popoli *et non vadi mendicando il viver.* Io per questo ho mantenuto per tre anni che non si sono ammazzati animali bovini."

molta prudentia terminato che più non si piantassero vigne, con un honestissimo temperamento, dichiarando, che quelli terreni nei quali non era stata vigna per dieci anni avanti e che vi potesse entrare aratro, che là non si debba più piantare vigne, ma lassarlo per il grano."

The just and honourable opinions expressed in these Reports of Garzoni and Foscarini do not appear to have been so acceptable to the Venetian Senate as the language held by Fra Paolo Sarpi, who addressed the Most Serene Republic, on the subject of her Greek Colonies and possessions, in the following terms[12]:

"For your Greek subjects of the island of Candia, and the other islands of the Levant, there is no doubt but there is some greater regard to be had of them, first, because that the Greek faith is never to be trusted; and perhaps they would not much stick at submitting to the Turk, having the example of all the rest of their nation before their eyes: these therefore must be watch'd with more attention, lest, like wild beasts, as they are, they should find an occasion to use their teeth and claws. The surest way is to keep good garrisons to awe them, and not use them to arms or musters, in hopes of being assisted by them in an extremity: for they will always shew ill inclinations proportionably to the strength they shall be masters of, they being of the nature of the gally-slaves, who, if they were well us'd, would return the kindness, by seizing the gally, and carry it and its commander to Algiers: wine and bastinadoes ought to be their share, and keep good nature for a better occasion.

As for the gentlemen of those Colonies, you must be very watchful of them; for besides the natural ferocity of the climate, they have the character of noblemen, which raises their spirits, as the frequent rebellions of Candia do sufficiently evidence. The use of Colonies was advantageous to the Roman Commonwealth, because they preserv'd even in Asia

[12] The opinion of Padre Paolo, &c. (delivered in 1615.) translated by Aglionby, Lond. 1689. See pp. 38—41. for the extract here given.

and Africa, Roman inclinations, and with them a kindness for their country. *If the gentlemen of these Colonies do tyrannize over the villages of their dominion, the best way is not to seem to see it, that there may be no kindness between them and their subjects;* but if they offend in any thing else, 'twill be well to chastise them severely, that they may not brag of any priviledges more than others: It will not be amiss likewise to dispute all their pretensions to any particular jurisdiction; and if at any time their nobility or title be disputed, it will do well to sell them the confirmation of it at as dear a rate as possible: and, in a word, remember that all the good that can come from them, is already obtain'd, which was to fix the Venetian dominion; and for the future there is nothing but mischief to be expected from them."

STATISTICAL APPENDIX.

CONTENTS.

EXPORTS.

Article.	Quantity.	Value in Turkish Piastres.	Duty.	Amount of Duty.
Oil................	3,500,000 okes.....	8,750,000	16 paras per oke...	1,400,000
Soap...............	50,000 quintals	6,500,000	11 piastres the quint.	550,000
Almonds..........	20,000 measures	140,000	3 per cent. and 20 paras the measure	14,200
Almonds shelled..	10,000 okes.....	55,000	3 per cent. and 5 paras the oke......	1,900
Cheese.............	50,000 okes.....	137,500	3 per cent. and 10 paras the oke......	16,625
Silk[1]...............	11,000 okes.....	1,320,000	3 piastres the oke and 3 per cent...	—
Carobs.............	40,000 quintals	320,000	20 paras the quintal	20,000
Vallonea..........	2,500 id.....	80,000	3 per cent. and mirí.............	10,000
Wax...............	15,000 okes.....	202,500	1 piastre the oke...	15,000
Honey.............	14,000 id......	42,000	6 paras the oke ...	2,100
Chesnuts..........	200,000 id......	100,000	3 paras the oke ...	15,000
Raisins[2]...........	6,000 quintals	168,000	3 per cent. and 30 paras the quintal	9,540
Wool..............	130,000 okes.....	390,000	4 paras the oke ...	13,000
Apples and Pears	150,000 id......	75,000	5 piastres for 1000	7,500
Oranges & Lemons	2,500,000 in number	200,000	4½ piastres for 1000	11,750
Linseed............	25,000 okes.....	15,620	3 per cent. and 2 paras the oke......	1,710
Indian Corn.......	30,000 id......	30,000	id. id. ...	2,400
Snails[3].............	20,000 id......	15,000	2 paras the oke ...	1,000
		18,540,620		2,091,725

[1] This exported silk ought to pay 72,600 piastres of duty, but no duty whatever is obtained, since all the silk is smuggled out of the island. The merchants of Megâlo-Kástron and Khaniá, themselves the exporters of it, are my authority for its amount.

[2] The exportation of raisins, before the vineyards were destroyed by the revolution, used to amount to 60,000, 80,000, and even to 100,000 quintals annually; and, were the country well governed, would soon again reach and exceed its old amount. On the exportation of both raisins and wine, from Crete, at the beginning of this century, see also above, p. 296.

[3] See the observations made above, p. 163. With respect to the third article mentioned in the opposite list of imports, it may be observed that every Cretan peasant wears *boots*, (see above, p. 253.) which cost him from 60 to 80 piastres: allowing that he uses one pair annually, and that 35 piastres will suffice for the shoes of his wife and family, we should obtain about 2,800,000 for the 25,000 families of the island. I mention this to confirm the truth of these items; for although they are obtained after a very careful investigation, and repeated conversations with all the principal merchants in the island, yet, when a country is possessed by Turks, every statistical enquiry is difficult, and its results need all the confirmation which independent calculations can give them.

IMPORTS.

Article.	Quantity.	Price in Turkish Piastres.	Value in Turkish Piastres.	Duty.	Amount of Duty.
Manufactured Goods[1]	—	—	3,630,000	3 per cent......	108,900
Woollen Cloths[1]....	400 pieces ...	400	160,000	id.	4,800
Buffalo Hides, other Skins, Morocco Leather, &c. &c.	—	—	2,800,000	id.	84,000
Wheat[2]	70,000 quilots..	14	980,000	Free.	—
Barley[2]	360,000 id......	6	2,160,000	Free.	—
Beans, Peas, Chicory, &c.	90,000 id......	12	1,080,000	Free.	—
Rice[3]	280,000 okes.....	2	560,000	Free.	—
Soda (for making soap)	35,000 quintals	45	1,575,000	3 per cent......	47,250
Cod Fish[4]	2,500 id......	100	250,000	id.	7,500
Other Salt Fish......	3,500 id......	150	525,000	id.	15,750
Tobacco..............	160,000 okes.....	4	640,000	5 paras the oke	20,000
Snuff	2,000 id......	25	50,000	1 piastre the oke	2,000
Coffee.................	50,000 id......	8	400,000	5 paras the oke	6,250
Sugar..................	80,000 id......	5	400,000	3 per cent......	12,000
Rum[5]...............	250 cases.... 20 barrels..	—	57,500	Farmed.	—
Wine.................	5,000 id......	50	250,000	Farmed.	—
Rakí..................	25,000 okes.....	3	75,000	Farmed.	
Wood.................	60,000 planks...	4	240,000	3 per cent......	7,200
Ditto from Constantinople........	10 cargoes..	15,000	150,000	id.	4,500
Firewood[6]	60 id......	5,000	300,000	Free	
Butter[7]...............	50,000 okes.....	7	350,000	3 per cent......	10,500
Oxen	400 head.....	200	80,000	id..........	2,400
Sheep.................	5,000 id......	30	150,000	id..........	4,500
Drugs and Groceries..............	15,000 okes....	mean price 10	150,000	id..........	4,500
English Tin.........	30 barrels...	1,500	45,000	id..........	1,350
Lead..................	10,000 okes.....	2	20,000	id..........	600
Nails..................	200 barrels..	200	40,000	id..........	1,200
Bar Iron	1,500 quintals	60	90,000	id..........	2,700
Cutlery[8]	—	—	300,000	id..........	9,000
Steel..................	100 cases.....	180	18,000	id..........	540
Paper..................	1,500 reams ...	15	22,500	id..........	675
Pottery...............	20,000 dozen....	6	180,000	id..........	5,400
Glass..................	100 cases.....	500	50,000	id..........	1,500
Salt....................	20,000 measures	2	40,000	20 paras mirí	10,000
			17,818,000		375,015

1 Cotton twist, calicoes, silk stuffs, linen, &c. It must be remembered that these manufactures serve to clothe a population of more than 25,000 families; so that although the sum seems considerable, yet it only amounts to about 150 piastres for each family. The calicoes and cotton goods in general come from England, by way of Sýra; most of the silks are of Turkish manufacture. The woollen cloths pass from the manufactories of Belgium, to Trieste, and thence to Crete.

2 Before the revolution the island, though its population then amounted to near 300,000, *exported* wheat annually, and sufficient wheat and barley remained for the home consumption.

3 Rice, though free from duty on importation, is liable to a duty of one para per oke, if re-shipped for any other part of the island. This duty has been imposed by the Egyptians.

4 This codfish is brought by French ships. The French government, desirous of encouraging the cod-fishery, gives a considerable bounty on the exportation of the fish from France.

5 Leghorn and Smyrna are the ports whence the rum is usually brought to the island.

6 For the soap manufactories. This firewood is from the forests of Karamania, Anatolia, and Rumelia.

7 The butter is usually imported from Barbary; a small quantity is sometimes brought from Russia.

8 Trieste is the great mart whence Crete, as well as most of the Levant, is supplied with cutlery, paper, pottery, glass, &c. of German manufacture, far inferior in quality, and at somewhat lower prices than the ordinary produce of English art.

BRANCHES OF REVENUE FARMED BY CERTAIN CONTRACTORS.

Article.	Megálo Kástron.	Rhithymnos.	Khaniá.	Total in Turkish Piastres.
Butcher's Meat[1]	57,000	9,000	14,800	80,800
Tobacco[2]	42,500	25,000	50,000	117,500
Snuff	8,000	3,000	6,000	17,000
Roasted Coffee	18,000	2,200	12,650	32,850
Fresh Fish	5,000	435	2,500	7,935
Wine and Spirit	55,000	15,000	36,060	106,060
Hides and Leather	50,000	47,000	66,100	163,600
Porters[3]	12,000	2,500	4,800	19,300
Oil Porters[4]	....	....	2,400	2,400
Wax	6,000	1,000	7,050	14,050
Octroi on all articles of consumption which enter the city	60,000	2,500	4,560	67,060
Duty on the hire and sale of horses[5]	7,000	....	6,996	13,996
Dyers[6]	16,000	4,500	6,600	27,100
Salt	26,400	....	20,100	46,500
The Weighers[7]	....	2,800	....	2,800
				718,951

[1] In each of the three principal towns the contractor pays the government the sum mentioned, and receives, at the gate of the town, ten piastres for every bullock, one piastre for every sheep, &c. as they enter.

[2] The contractor collects and retains the duty of one piastre on each oke of tobacco, paying annually to the government the sums mentioned. So of the other items.

[3] Here those who exercise the calling of porter have to give one tenth of all their earnings to the government contractor.

[4] Probably there is a separate contractor for the oil-porters at Megálo-Kástron and Rhithymnos, as well as at Khaniá.

[5] I am not sure of the accuracy of this sum for Megálo-Kástron: at Khaniá the amount annually paid by the contractor is exactly 6,996 piastres, as mentioned.

[6] The dyers can only purchase their drugs of these contractors, who alone can import the drugs required for the art of dyeing into the island.

[7] I know not whether the office of weighers be paid for at the other two towns. The farmer or his deputy, by virtue of this purchased privilege, weighs articles of exportation on the quay, and receives a certain small payment for his trouble.

SEVENTHS RECEIVED BY THE GOVERNMENT IN CRETE.

One of the Consular Agents at Khaniá furnished me with what he called a Table of the Revenues of Crete: but the document was so wide of the truth, in almost all its details, that I only made use of it to shew him how imperfect had been his estimates, and how erroneous were his calculations. Since, therefore, my Table is solely the result of my own investigations, I will explain the data on which the determination of the amount of the principal items depends.

The most important article of produce and exportation is oil: the amount exported differs greatly in different years, the crop of olives being very variable: but all persons acquainted with the subject agree that the mean annual exportation, from all the ports of the island, amounts to not less than 3,500,000 okes, and that about 50,000 quintals of soap are also exported; so that if, to simplify the question, we state both soap and oil under the latter title, we shall have an annual exportation of about 5,000,000 okes of oil.

If we can ascertain the home consumption of oil we shall plainly know the whole produce.

Now there are about 26,000 families in the island, and the result of inquiries, more than once alluded to in the previous pages, as to the annual consumption of each house was always the same: every body asserting that each Cretan family consumes *at least* ten to twelve mistata, or about one hundred okes annually. Wherever any one's condition is at all superior to that of a common peasant, the consumption is much greater. Hence, therefore, the whole home consumption amounts to about 2,600,000 okes, and, with the exported oil and soap, gives 7,600,000 okes as the mean annual produce, after the payment of the government's seventh, taken always in kind and usually sent to Alexandria, which seventh consequently will be,

		Okes.
		1,266,666
Home consumption	2,600,000	
Exported	5,000,000	
		7,600,000
Whole produce		8,866,666

	Piastres.
Now these 1,266,666 okes of seventh are worth	3,166,665
And the seventh (τὸ ἕβδομον) of all other articles of produce, such as corn, carobs, almonds, wax, honey, silk, &c. is, in *every part* of the island, equal to that of oil, and in *most parts* exceeds it very considerably, and may therefore safely be put down at	4,333,335
Whence the total amount of the seventh is	7,500,000
The Viceroy, however, not having yet deprived all the owners of Mukatás of their tithes, we must deduct, as received by those who still enjoy their old fiefs	1,100,000
And we, therefore, have as the amount of Territorial Revenue annually received by the government	6,400,000

As a confirmation of these calculations I may mention that on learning from certain officers, who are entrusted with the collecting of this seventh of every kind of produce for certain cantons, its amount for their respective districts, and making these data the basis of an independent calculation of the amount of tithe obtained by the government from the whole island, I obtained a result so nearly agreeing with this as fully to establish the accuracy of my calculations.

BALANCE SHEET OF RECEIPTS AND EXPENSES

RECEIPTS.

	Piastres.
Tithe (the seventh) of all produce	6,400,000
Duty on sheep and goats	200,000
Duty on all the exports[1] (oil, soap, cheese, wax, carobs, &c.) ...	2,091,725
Import duties	375,015
Branches of revenue farmed out to contractors, octroi, &c.	718,951
	9,785,691

Thus the net revenue of the island is a little more than 100,000*l.*, and is therefore greater than would be the amount of any reasonable expenses of governing it, whether by Mehmét-Alí or by a European power, as England.

[1] It is probable that the government may not receive quite so much as these 2,091,825 piastres, for the merchants themselves admit that a great deal, even of oil, is exported without paying the duty. This is only a natural consequence of the very low salaries paid to all the custom-house officers, who are consequently easily bribed.

OF THE GOVERNMENT IN CRETE.

EXPENSES.

	Piastres.
Salary of the Governor General[1]	800,000
Expenses of the Councils, viz. at Megálo-Kástron, Khaniá, Rhíthymnos, and Sfakiá	535,540
Treasury	120,000
Custom-house department	49,744
Government printing-office	27,335
Old Turkish court of justice	24,000
Regular Arab troops—1 general, 2 colonels, 2 lieut.-colonels, &c. &c. 384 corporals, and 3,840 privates, band of music, instructors, &c. &c.	2,464,040
5,500 daily rations	2,007,500
Arms, accoutrements, and the expense of keeping some mules	1,630,000
The Albanians—(1,500 irregular troops, some mounted,) cost annually about	1,760,000
	9,418,159

[1] Mustafá-pashá's salary is nominally 2,500,000 piastres: but Méhmet-Alí takes the rents of all his estates in Egypt, (formerly the property of his uncle Khassán-pasha,) and, *as governor of Crete*, he cannot be considered as receiving more than the sum put down, between 8,000*l.* and 9,000*l.*, which exceeds the salaries of most governors in British possessions.

THE

VILLAGES AND POPULATION

OF THE SEVERAL

DISTRICTS OF CRETE

In 1834.

* *The letters* C. F. *mean "Christian families," and* M. F. *"Mohammedan families."*

KISAMOS.

	C. F.	M. F.
Polemárkhi	38	4
Rhodhopú	96	3
Skutelónas	60	0
Afráta	20	0
Rhavdhúka	20	0
Kamarakumúli	18	2
Nokiá	28	0
Spélaea	60	1
Vuvés	28	2
Múleti	30	2
Ghavalomúri	20	2
Vukoniés	0	50
Anoskelí	50	0
Karéa	0	18
Aspriliáno	0	15
Stazipragú	3	20
Rúmata	35	35
Kakópetro	4	4
Panéthimos	30	10
Maláthero	20	5
Perivolákia	0	20
	560	193

	C. F.	M. F.
Nopía	4	0
Kotsianá	0	10
Dhrapaniá	30	0
Potamídha	50	10
Vurgháro	20	8
Murí	30	2
Topoliá	50	2
Hághios Kyr Iánnes[1]	15	2
Kaláthines	30	2
Kalerianá	26	5
Kurfalónas	6	2
Kísamo-Kastéli	0	30
Palaeókastro	35	5
Priniákos and Lusákies	40	6
Mesághia[2]	216	15
Sirikári	60	0
Enneá-Khoriá	286	45
	898	144
	560	193
Total	1458	337

[1] Called *Kyrghiani*, in Lapie's map and elsewhere.

[2] The villages of this district are Haghíus Pántas, Kamártzo, Hághio Gheórghi Kalérghi, Neokhorió, Kalyviané, Piperianá, Azogeraéa, Télipho, and Kutrí.

SELINO.

	C. F.	M. F.		C. F.	M. F.
Sklavopúla	38	0	Akhladhiákes	0	10
Kalámi	12	0	Stráti	0	12
Mustáko	6	2	Teménia	14	8
Khasí	5	35	Asfendílias	5	3
Hághio Thódhoros	12	4	Strovoliés	6	30
Pelekánas	6	10	Mylonés	0	8
Katsivelianá	2	3	Sasalos	4	18
Stavrós	4	5	Papadhianá *cum* Merté	2	2
Mesóruma	0	12			
Khádhros	0	0	Kamaliá	7	2
Sarakéna	0	20	Rhodhováni	30	1
Kontokynéghi	0	1	Moné	4	12
Tsalianá	3	0	Livadhá	30	0
Vasiláki	3	0	Krustoghérako	20	0
Kálamos	0	12	Agrilé	18	6
Spaniáko	2	18	Livádhes	8	7
Vlithiás	8	3	Marália	17	2
Makhiá	10	3	Kabanú	37	4
Kakotíkhi	15	20	Skáfi, Ergastéri	27	10
Sfakó	12	2			
Gligorianá	5	3	Lukianá	2	6
Kopetí	1	5	Tzagariáko	0	10
Dhryes	0	16	Tzikianá	7	1
Psarogharianá	0	3	Prínes	22	7
Alighí	0	12	Epáno-khorió	30	1
Kántanos[1]	6	70	Haghía Iréne	50	2
Spína	10	0			
Floriá	0	15		340	162
Azoeré	0	15		160	289
	160	289	Total	500	451

[1] A district containing the villages of Kufalatós, Plemmenianá, Trígonos, Trakiniákos, Aniseráki, and Babakádhos.

KHANIOTIKA.

	C. F.	M. F.
Keramia, containing Kontópulo, Gheroláko, Dhrakóna, Speliariá, and *Papadhianá*[1]	194	0
Thériso	80	2
Mesklá	72	3
Lâki	140	2
Şévronas		
Néa Rúmata		
Prasés		
Skoedhiá		
Khosté	150	12
Orthúni		
Karés		
Kháranos		
Skafidhákia		
Kámpi	0	0
Sirilé	10	6
Mése Sirilé	10	6
Néo-khorió	3	2
Vlakeronétissa	14	2
Pyrgo-Psylónero	10	1
Xeniko-khóri	2	1
Psylles	10	0
Kontomarí	3	0
Derés	4	3
Skonízos	12	2
Vatólako	50	15
Alikianú	40	9
Kúfo	6	4
	810	70

	C. F.	M. F.
Furnés	30	6
Skénes	20	4
Papiólos	6	3
Apothékes	7	0
Manoilótos	14	3
Psathoghiáno	14	12
Lutráki	6	2
Gheráni	16	8
Módi	18	18
Pateléri	6	4
Platanias	40	4
Galata	70	10
Episkopé	0	1
Haghía Marína	20	8
Daratzús	20	5
Varypetro	18	0
Stalo	12	0
Katsistro-khóri	4	0
Perivólia	30	6
Murniés	70	8
Tzikalariá	40	0
Nerokúru	30	6
Maláxa	25	0
Zurva	15	0
Vrysis	20	10
Kertomádho	6	5
Haghiá	8	4
Akroteri	250	45
Omissions	30	10
	845	182
	810	70
Total	1655	252

[1] Papadhianá again contains Akhládhes, Hághio Gheórghi, Lúlos, Aletrivári, Panaghiá, Khorafianá.

SFAKIA.

	C. F.	M. F.		C. F.	M. F.
Samaría	10	0	Vraska [3]	0	0
Haghía Ruméli . .	38	0	Vuvas	1	0
Hághio Iánnes . .	40	0	Asfénto	55	0
Arádhena	36	0	Anomikianá [4] . . .	1	0
Livanianá	20	0	Kolokásia [4]	0	0
Lutró [1]	3	0	Kalikráti	85	0
Anópolis	115	0	Patsianó [5]	14	0
Murí	50	0	Askýfo	160	0
Sfakiá [2]	60	0			
Komitádhes	80	0		316	0
Nípros	80	0		532	0
	532	0	Total . . .	848	0

The number of Sfakians capable of bearing arms at the present day is about 1200. There are nearly 200 Sfakian families now settled out of Crete, at Mélo and elsewhere. The number of souls in Sfakiá before the year 1821 was about 12,000. Maniás and others say, that they might be considered as forming three bodies, one of which lived by their flocks: another third were sailors, the port of Lutró and also that of Sfakiá possessing several vessels, those of the former being large enough to trade to Smyrna and Alexandria. The third division were πραγματευτάδες, or travelling merchants (pedlars) who used to hawk their goods all over the island.

[1] That is, three permanent residents: Lutró is the winter-village of Anópolis (see above, p. 242.) and during the winter-months its population is much the same as that of Anópolis in the summer.

[2] Sfakiá is the chief winter-residence of the Askyfiótes: a few families go elsewhere, as to Dhrámia: see above, Vol. I. p. 73.

[3] The winter-village of Nípro.

[4] Winter-village of Asfénto.

[5] Winter-village of Kalikráti.

APOKORONA.

	C. F.	M. F.		C. F.	M. F.
Karés	30	1	Kástelos	20	0
Rhamné	25	3	Pátema	15	2
Melidhóni	20	0	Flakí	25	4
Khiliomudú	9	0	Goniá	65	0
Samónas	7	0	Dhrámia	(Sfakians)	
Kyriakusália	6	0	Xópolis	10	6
Stylo	5	1	Azoeromúri	2	6
Próverma	16	2	Katuna	10	0
Makérus	30	2	Kalamítzi	45	4
Neo-khorió	50	3	Karydhi	6	3
Nero-khóri	9	3	Vámos	30	15
Paidho-khóri	10	5	Arménus	60	15
Pemónia	25	10	Varás	16	4
Fré	80	1	Kalyves	50	20
Kúkos	6	0	Dulianá	10	0
Sifianá	4	0	Kavallo-khóri	10	20
Tzifés	40	3	Koprána	5	6
'Ipos	70	4	Hághio Vasíli	2	5
Prósnero	75	15	Kokkino-khorió	40	0
Vaffé	65	3	Pláka	20	0
Alíkampo	75	0	Dhrápanos }		
Phílippos	15	0	Kefalás }		
Máza	15	0	Xerostérni }		
Khábatha	6	0	Vuri }	300	25
Phoné	18	0	Seliá }		
Apriskia	15	0	Likotinará }		
Mathés	15	0			
Murí	10	2		741	135
Kurna	70	4		821	62
	821	62	Total	1562	197

HAGHIO VASILI, OR LAMPE.

	C. F.	M. F.
Kapsodhásos . . .	30	0
Skaloté	30	0
Argúles	15	0
Rhodhákino . . .	45	0
Phoenikiás[1]	6	0
Seliá	40	0
Myrthió	30	0
Mariá	14	0
Asómatos (village) .	15	0
Levkoïa	30	0
Ianniú	15	0
Préveli	5	0
Keramé	40	0
Vrímisko	5	5
Váto	25	0
'Ardhakto	25	0
Kissós	15	15
Total . . .	385	20

	C. F.	M. F.
Vrysis		
Akúmia		
Dumaghérgo . .		
Platané		
Krya-Vrysis . . .		
Mélabes		
Aktúdha		
Saktúria	388	0
Murn		
Daravianá . . .		
Mixóryma . . .		
Haghía Pelagía .		
Hághio Vasíli . .		
Hághio Iánni . .		
Kalé Sykiá . . .		
Speli		
Labiné		
Koxaré		
Láko		
Diblokhóri . . .		
Vratí	50	188
Korédho		
Atspadhiá . . .		
Tzikalariá . . .		
Palaeolutrá . . .		
Kuselianá		
	438	188
	380	20
Total . . .	818	208

[1] This hamlet is little more than a metókhi of the monastery of Prophét-Elías at Rústika, on which see Vol. I. pp. 97—99.

RHITHYMNIOTIKA.

	C. F.	M. F.
Episkopé . . . . :	35	25
Arkhúdhona . . .	30	30
Pólis	95	2
Hághio Gheórghios	20	6
Hághio Konstantíno	40	10
Rústika	50	10
Palaelímno	0	25
Kalonykti	6	5
Epáno-Vasermónero	0	25
Káto-Vasermónero .	0	30
Hághios Andhréas	4	20
Koniá	25	4
Metókhia	24	0
Iarrán	30	5
Priné and Alitsópulo . . .	40	90
	399	287

	C. F.	M. F.
Peghé	38	2
Haghía Paraskevé	8	3
Marulá	65	5
Bagalokhóri	10	5
Khamalévri	10	5
Arsáni (monastery)	2	0
Hághio Iánni . Haghía Kyriaké Mése Kyrianá . Pikré, Amnáto . Lútra, Adhíli .	50	180
Gheráni and other villages omitted, about	60	65
	243	265
	399	287
Total . . .	642	552

AMARI.

Amári	20	1
Mérona	54	6
Edánoso	8	7
Patsó	0	20
Karínes	4	3
Ghenní	2	4
Gerakéres	20	20
Elaínes	6	4
Gurgúthos	12	0
Tze Vrysis	25	0
Smilé	1	5
Tze Dhryés . . .	5	2
Anómeros	50	0
	207	72

Khordháki	0	8
Hághio Iánni . . .	15	0
Apodhúlo	14	6
Haghía Paraskevé .	15	0
Vathiakó	0	15
Sáhta	1	6
Klíma	20	0
Lokhrié	6	0
Ardhakto	0	10
Plátanos	15	20
Níthavri	13	2
	99	67
	207	72
	306	139

	C. F.	M. F.
Kurútes	5	16
Fúrfuras	40	6
Visári	20	0
Plataniá	10	6
Bístaghi	50	2
Kaloghéru	7	0
Thrónos	9	6
Klisídhi	0	10
	141	46
	306	139
	447	185

	C. F.	M. F.
'Opsighas	7	1
Monasteráki	20	0
Laviótes	8	0
Aposéti	12	1
Ghénna	0	20
Apostólus	3	4
Voliónes	0	35
	50	61
	447	185
Total	497	246

MYLOPOTAMO.

I have mislaid my list of villages of Mylopótamo, and cannot lay my hand on it while this sheet is passing through the press. The principal village is Margarítes, with a population of a hundred Christian families. The village Tripódhos (see Vol. I. p. 314.) is in Mylopótamo, near Elévtherna. The whole population of the eparkhía is, C. F. 920. M. F. 310.

KASTEL-PRIOTISSA.

	C. F.	M. F.
Hághio Iánni	8	3
Khameléri	30	0
Itsídhia	30	0
Síva	22	0
Kalývia	0	15
Vórus	25	5
Phaneroméne	10	10
Kissús	0	14
Kalokhorafí	4	4
	129	51

	C. F.	M. F.
Megarikhári	25	2
Gligoriá	3	18
Temenéli	0	5
Kamáres	8	8
Lagolió	2	4
Dibáki	95	3
Hághia Triádha	6	0
	139	40
	129	51
Total	268	91

KAENURIO KASTELI.

	C. F.	M. F.		C. F.	M. F.
Bobia	50	0	Moróni	0	30
Petrokefáli	12	5	Kirmusí	0	10
Khusé	4	0	Makrianá	3	0
Lístaro	6	0	Paliamo	6	0
Alithini	3	12	Panaghiá	5	0
Péri	4	0	Kúrtes	0	20
Akustulianá	12	0	Vorízo	35	0
Plóra	18	3	Zaró	50	6
Anoía	5	15	Zérzeri	41	4
Plátanos	15	0	Panasú	6	12
Lakiákes	2	1	Múlia	5	10
Agavalianá	3	0	Priniá	15	2
Pigaitháki	6	1	Ráfti	0	18
Manusaná	2	0	Varvulíti	0	12
Myres	6	6	Vasiliké	5	10
Kaparianá	0	7	Kadhílo	5	0
Siferianá	0	10	Króta	8	0
Talia	0	15	Miamú	15	0
Monókhoro	3	0	Diskári	12	0
Metrópolis	5	5	Monastéri	3	0
Vreli	1	5	Nívreto	5	5
Apólykhnos	0	8	Skúrvula	3	0
Hághii Dhéka	10	10	Omitted, a few metókhis,	12	8
Ampelússa	5	6			
Rufá	0	10			
Plutó	0	12		234	147
				172	131
	172	131	Total	406	278

MONOFATSI.

Megále Vrysis	25	2	Haghía Varvára	32	8
Pervelianá	10	0	Grándhos	0	5
Hághio Thómas	20	1	Katomuliá	0	5
Argáthia	8	4			
Axéndhi	0	30		32	18
				63	37
	63	37		95	55

	C. F.	M. F.
Vurió	0	5
Valí	0	10
Xerókampos	0	2
Gangális	0	24
Stólus	0	6
Riziká	0	4
Oeniá	0	8
Laráni	0	20
Atzipádhes	2	8
Akria	0	10
Lúres	0	8
Asémi	8	7
Sokará	2	38
Apoénia	0	5
Velúdhi	0	2
Protória	12	8
Dhamántri	0	3
Plakiótissa	0	2
Iligórtyno	2	15
Teféli	0	30
Karáki	0	6
Voreá	0	10
Kutú	0	2
Futikastéli	0	8
Kakó-khorió [1]	0	20
Hághio Gheórghi Epáno-Siphes (monastery)	2	0
Madhé	0	10
Aposeléni	0	6
Armanóïa	0	5
Kefála	0	8
Dhamánia	0	12
	28	302
	95	55
	123	357

	C. F.	M. F.
Arkâdhi	0	6
Kólona	0	2
Dhúli	0	15
Méso-khorió	6	9
Fléga	0	2
Rotás	0	20
Pyrgo	2	30
Dhukári	1	25
Kháraka	25	10
Haghía Photía	0	2
Akéntria	0	20
Ethiá	10	5
Murniá	0	1
Priniá	2	0
Amygdhalo	0	12
Paranyphus	5	1
Platánia	0	4
Kapitanianá	30	0
Kalálaga	4	0
Zábres	0	2
Stérnes	5	4
Panaghía	0	12
Dhionysia	0	20
Kária	0	4
Akámuto	0	4
Avgés	4	4
Vagoniá	6	8
Gumása	5	0
Lúkia	8	0
Selámi	1	0
Furnofárago	1	5
Six other villages [2]	0	1
	115	228
	123	357
Total	238	585

[1] See Vol. I. p. 82.

[2] Five of which have not had a single inhabitant in them since they were destroyed by the Christians during the war.

MALEVIZI.

	C. F.	M. F.		C. F.	M. F.
Akhládha	40	0	Sárkos	20	0
Rhogdhiá	50	0	Kitharídha	15	0
Tylisso	90	10	Khrusónas	140	0
Moné	30	5	Káto-Asítes	30	0
Kamári	30	10	Epán-Asítes . . .	25	3
Korufés	30	0	Priniá	15	6
Keramútzi	0	20	Syva	22	3
Pedamódhi	5	15	Kerásia	25	4
Vútes	30	0	Avgenikí	35	2
Kalésia	25	0	Dháfnes[1]	35	2
Stavrákia	30	0	Goniés	40	0
Petrokéfalo	14	12		402	20
Hághio Myro . . .	100	2		499	74
Pyrgu	25	0			
	499	74	Total . . .	901	94

TEMENOS.

Dháfnes	20	0	Sílamos	10	0
Veneráto	25	0	Ten or twelve other hamlets . .	16	94
Kaní-Kastéli . . .	28	18			
Arkhánes	160	6		26	94
Kát-arkhánes . . .	30	0		263	24
	263	24	Total . . .	289	118

[1] Only part of the village is in Malevízi.

PEDHIADHA.

	C. F.	M. F.
Karteró	9	7
Skalára	10	6
Patsídhi	4	6
Kunávi	10	10
Katalagári	16	0
Kudhétsi	20	5
Hághio Vasíli	15	10
Skylus	10	12
Melísses	16	4
Haghías Paraskés	12	12
Sgurokephâli	25	0
Episkopé	20	15
Haghía Tanía	25	0
Stamniú	10	0
Elaiá	20	16
Vathiá	0	10
Kato-Vathiá	10	3
Kaenúrio-khorió	20	20
Ghálipi	15	10
Ghalífa	30	0
Kharasó	15	0
Smári	25	0
Apostólus	16	16
Zophórus	10	10
Trapsanó	38	0
Rússo-khorió	0	12
Voni	0	30
Muktárus	30	3
Varváro	3	10
Varvárus	10	12
Sklaverokhóri[1]	5	1
	449	240

	C. F.	M. F.
Lulianó	3	1
Kastéli	18	15
Dhiavaidhé	4	6
Peghaidhúri	5	4
3 Metókhis	25	0
Xydhá	20	0
Kastamonítsa	35	0
Amarianó	15	0
Mathiá	17	0
Haghía Paraské	6	0
'Ipito	15	0
Gheráki	24	0
Epídheikto	10	0
Navlí	0	12
Kasámi	5	5
Panaghía	15	15
'Ebaro	10	15
Askús	20	2
Avdhú	75	6
Goniés	38	0
Gherá	18	0
Krási	60	0
Mokhó	120	1
Mália	100	0
Khersónesos (with the two other neighbouring villages)	60	3
Stravorína	20	0
	738	85
	449	240
	1187	325

[1] This and a few of the preceding names seem to indicate a Slavonian immigration into this part of the island, which is in the immediate neighbourhood of Khersónesos and Lýttos, the locality in which an earlier and more important colony established itself: see Plutarch, de virt. mul. Tom. ii. p. 247. d.

	C. F.	M. F.		C. F.	M. F.
Kóxari	8	5	Philíssa	6	5
Vurú	3	1	Pezá	5	5
Skoteinó	8	5	Skótola	0	15
Gúves	35	10	Gaidhuriáni	3	6
Anópolis	25	0	Komés	2	4
Potamiés	10	10	Sphendíli	8	0
Keliá	5	5	Kerá	25	0
	94	36		49	35
	1187	325		1281	361
	1281	361	Total . . .	1330	396

RHIZO-KASTRON.

Khristós	38	1	Fabrianá :	0	4
Parsá	29	0	Phílippo	0	8
Epáno-Sími	96	0	Kalyvia	5	3
Káto-Sími	38	10	Akhendhriá	6	0
Kalamí	44	0	Kakú, ...	0	11
Sykológo	37	0	Tourlotí	0	5
Epáno-Pévko } Káto-Pévko } .	56	0	Katsikali	0	4
			Gharípa	0	12
Kéfalo-vrysi	18	8	Káto-Gharípa . . .	0	4
Myrá	62	1	Pyrathi	0	3
Vakhó	10	1	Kamáres	0	1
Hághio Vasíli . . .	34	7	Amurghiéles	0	9
Virevatá	8	2	Gurniá	0	6
Viáno	103	90	Sitironá	0	4
Káto-Viáno	0	16	Nispitá	0	4
Khondhró	10	40	Bathiá . :	1	13
Mesé	3	1	Partirá	0	22
Skhoeniá	11	0	Vitsiliá	0	6
Dhemáti	11	1	Haghía Simé . .	0	9
Kastelianá (with the neighbouring metókhis)	11	0	Epáno-Myliárisi . .	1	4
				13	132
				619	178
	619	178		632	310

	C. F.	M. F.		C. F.	M. F.
Káto-Myliárisi . .	0	2	Hághii-Apóstoli . .	0	0
Vutufú	5	6	Alagni	6	7
Puliá	0	16	Monasteráki . . .	0	6
Nusúta	1	12	Igni	0	5
Arkhalokhóri . . .	7	10	Makherá	0	13
Ghasi	0	8	Mitsitséri	0	4
Kuméri	3	5	Sorokhíano	1	4
Zidá	1	13	Vakhiotes	0	4
Patsídhero	7	4	Laguta	0	14
Rúma	0	0	Kefalá	0	0
	24	76		7	57
	632	310		656	386
	656	386	Total . . .	663	443

MIRABELLO.

Mílato	35	0	Skoeniá	30	0
Latsídha	35	5	Furné-kastéli . . .	50	0
Vulisméne	50	15	Káto-Furné	40	0
Kaenúrio-khorió . .	100	2	Epáno-Furné . . .	30	0
Neokhorió	20	20	Four monasteries .	35	0
Vrakhási . . / Vrysis . . . / Kommeriakó / Platy-Pódhi . / Nikithianó . / Lémnes . . .	80	80	Omissions . . .	50	30
			LASITHI.		
			Gaidhurómadra and Gerontomurí	30	0
			Pláti	25	0
			Psykhró	35	0
Krítsa	180	2	Maghulá	25	0
Krusta	30	0	Kamináki	30	0
Prina	40	0	Avrakódi . . . / Kudhumaliá . . . / Platianó. . . . . / Hághio Gheórghi	100	0
Kaló-khorió	10	0			
Alúdha	40	0			
Spína Lónga . . .	0	81			
Vruká	30	0		480	30
Lúma	15	0		665	205
	665	205		1145	235

	C. F.	M. F.
Hághio Kostantíno	30	0
Mésa Lasitháki . . / Mésa Lasíthi . . .	30	0
Monastery of the Panaghía Marmakéto	25	0
	85	0
	1145	235
	1230	235

	C. F.	M. F.
Farsáros and Dzermiádha	120	0
Lagú with Pinikianó	20	0
Potámus[1]	20	0
	160	0
	1230	235
Total . . .	1390	235

SETIA.

	C. F.	M. F.
Orno	30	4
Stavrodhoxári . . .	25	4
Lápitho	5	0
Graeés	15	2
Tornó	16	0
Pévkus	20	1
Dháfne	4	15
Hághio Mámas . .	5	3
Rhukáka	18	14
Krýa	7	3
Iskhiá	5	10
Lithínes	15	2
Vuriá	5	3
Papá-jannádho . .	8	0
Lutráki	0	5
Nethiá	2	10
Arméni	20	6
Kandhrá	20	7
Voilá	1	5
	221	94

	C. F.	M. F.
Zíro	30	30
Apídhi	10	0
Léthi	3	0
Tzo	25	0
Kaló-khorió	15	0
'Aspro-neró	3	5
Kamétulos	5	0
Lamnóni	0	8
Zakro	20	0
Adhravásto	10	0
Klisídhi	5	0
Azokéramo	8	0
Kelaria	10	0
Magasiá	75	8
Xerolímni	10	0
	229	51
	221	94
	450	145

[1] This village is not on the edge of the great plain of Lasíthi, but just over one κορυφή, as the Lasithiótes told me.

	C. F.	M. F.		C. F.	M. F.
Mesérios	10	0	Arnéku-Metókhia	22	0
Khónos	3	6	Haghía Phothía	5	0
Karýdhi	15	1	Episkopé (a second)	0	8
Sítanos	15	2	Piskokéfalo	26	14
Spélaea	8	0	Russiá-Ecclesiá	20	0
Kalamávki	0	10	Akhládhia	10	5
Katsidhóni	10	0	Paraspóri	0	15
Sandali	1	6	Skopé	18	3
Sfakiá	0	14	Kamézi	20	4
Vavélus[1]	0	25	Metokhiá	22	2
Kanénes	12	0	Turloté	37	0
Sklávus	8	2	Sfáka	32	7
Turtúli	5	10	Lástro	27	6
Episkopé	2	10	Káto (or Péra) Mulianá	60	8
Sotéra	2	3			
Marónia	8	0	Mésa Mulianá	50	3
Zú	0	9	Palaepétsi	0	6
	99	98		349	81
	450	145		549	243
	549	243	Total	898	324

HIERAPETRA.

	C. & M. F.		C. & M. F.
Anatolé	60	Kapístri	15
Malés	200	Vasiliké	25
Kalamávka	100	Kavúsi	100
Meseléri	40	Monasteráki	10
'Istronas[2]	20		150
Makriliá	10		430
	430		580

[1] This place had 150 tuféks, nearly six times its present number, at the outbreaking of the Greek insurrection in 1821.

[2] Also called Kaló-khorió.

	C. & M. F.		C. & M. F.
Epáno-khorió	50	Hághio Iánnis	40
Káto-khorió	70	Iannítsi	10
Episkopé	25	Kendhrí	35
Papadhianá	15	The Kastéli	300
	160		385
	580		740
	740	Total . . .	1125

Of these 1125 houses there are in the Kastéli 200 belonging to Mohammedans, and in the villages the number of Mohammedan families is about 85, which gives 840 as the number of Christian families of the district. No doubt this is very correct, the kharatjis of the district being exactly 900.

SUMMARY OF THE POPULATION OF CRETE IN 1834.

Name of District.	Chief Place.	C. F.	M. F.	Total.
1. Kísamos	Kísamo-kastéli	1458	337	1795
2. Sélino	Rhodhováni	500	451	951
3. Khaniotika	Khaniá	1655	252	1907
4. Sfakiá	Askyfo, Anopólis, Sfakiá	848	—	848
5. Apokórona	Fré, Kefalá	1562	197	1759
6. Hághio Vasíli	Mélabes	818	208	1026
7. Rhithymniotiká	Rhíthymnos	652	552	1204
8. Amári	Anómeros	497	246	743
9. Mylopótamo	Margarítes	920	310	1230
10. Kastel-Priótissa (or Mesará)	Ampelússa	268	91	359
11. Kaenurio-kastéli (or Mesará)		406	278	684
12. Monofátsi (or Mesará)		238	585	823
13. Malevízi	Hághio Mýro.	901	94	995
14. Témenos	Arkhánes	289	118	407
15. Pedhiádha	Mokhó, Mália	1330	396	1726
16. Rhizó-kastron	Viáno	663	443	1106
17. Mirabéllo	Kaenúrio-khorió	1390	235	1625
18. Setía	Mulianá	898	324	1222
19. Hierápetra	The Kastéli	840	285	1125
		16133	5402	21535

Therefore the rural population consists of about 81,000 Christians and 27,000 Mohammedans: a total of	108,000
Add the three chief cities—	
Megálo-Kástron	12,000
Rhíthymnos	3,200
Khaniá	5,800
So that the whole population of the island in 1834 amounts to about	129,000

This is about half its amount at the outbreaking of the Greek Revolution in 1821.

POPULATION OF CRETE AT DIFFERENT EPOCHS.

When, in the sixteenth century, the inhabitants of Crete had been reduced to about 250,000 souls, we find Proveditors complaining in their dispatches to the Senate at Venice, of the depopulation of the island, and speaking of whole tracts of ground, and, in some places, even of olive-trees, as left entirely neglected. An English traveller in Crete, a few years after the Mohammedan conquest, Randolph, estimates its reduced population at less that 80,000 souls. In spite of the fearful insecurity, for both person and property, which prevailed up to 1821, we find that the natural fertility of the country had reproduced, by that year, not less than 260,000 or 270,000 inhabitants, who were nearly equally divided between the two prevalent religions.

The fertility of Crete is such as warrants us to suppose, that in ancient times its population must have been somewhere near a million of souls. This amount would not people it so densely as either Malta or the adjacent island, called Gozzo by the Italians; in both of which the population is about 600 on the square mile, at the present day. The supposition of a million of inhabitants in *ancient* Crete, agrees very well with passages of old Venetian chroniclers, from which it may be inferred, that when the Venetians first attempted to acquire the island, at the beginning of the thirteenth century, its population was very much indeed greater than it was in the sixteenth. Probably the Venetians found 500,000 or 600,000 souls in the island. The century and a half of almost constant warfare, which followed their first landing, may well have produced a greater destruction of human life than has been effected by the few years of the recent war, which, as we see, have reduced the number of inhabitants to half their former amount.

ERRATA AND ADDENDA.

VOLUME I.

P. 11. lin. 15. for *Grece*, read *Greece*.
p. 298. 1. for *site; I*, read *site, even if I*.

VOLUME II.

p. 36. not. 21. A learned friend points out to me a passage of Herodotus, (VII. 37.) which it is extremely interesting to compare with the words of my ignorant Greek peasant.

p. 38. lin. 15. for *Mahommedans*, read *Mohammedans*.

86. 27. and 32. for *Kalamydes*, read *Kalamyde*.

221. 36. for *πρυΐ*, read *πρωΐ*.

285. 5. of not. 1. for *Racolta*, read *Raccolta*.

286. not. 2. (Compare Vol. I. p. 117. not. 51.) The *jus servile* forms a considerable title in all our old law-books, under the name of *villenage*. We had not only serfs or villeins regardant, who went with the land; but villeins in gross, who might be sold like any other chattels. (See ROLLE, Abr. II. 732. foll. Co. LITT. 116. 120. &c. DYER, Rep. p. 48. b.) The civil wars between the houses of York and Lancaster did much to extinguish this villenage, (as similar causes extinguished it in Italy much earlier). Nevertheless it prevailed in various parts of England under Elizabeth, and a case of it occurred as late as the reign of James I. It has been observed that, although the clergy tried to effect the enfranchisement of villeins belonging to the laity, yet the Church still retained its own serfs in bondage. "The holy fathers—had convinced the laity how dangerous a practice it was for one Christian man to hold another in bondage: so that temporal men—were glad to manumit all their villeins. But the said holy fathers, with the abbots and priors, did not in like sort by theirs; for they also had a scruple in conscience, to impoverish and despoil the Church;——and so kept their villeins still." SMITH's Commonwealth, III. 10. quoted by BLACKSTONE, Com. Vol. II. p. 96. Compare BARRINGTON, Observations on the more ancient Statutes, p. 277.

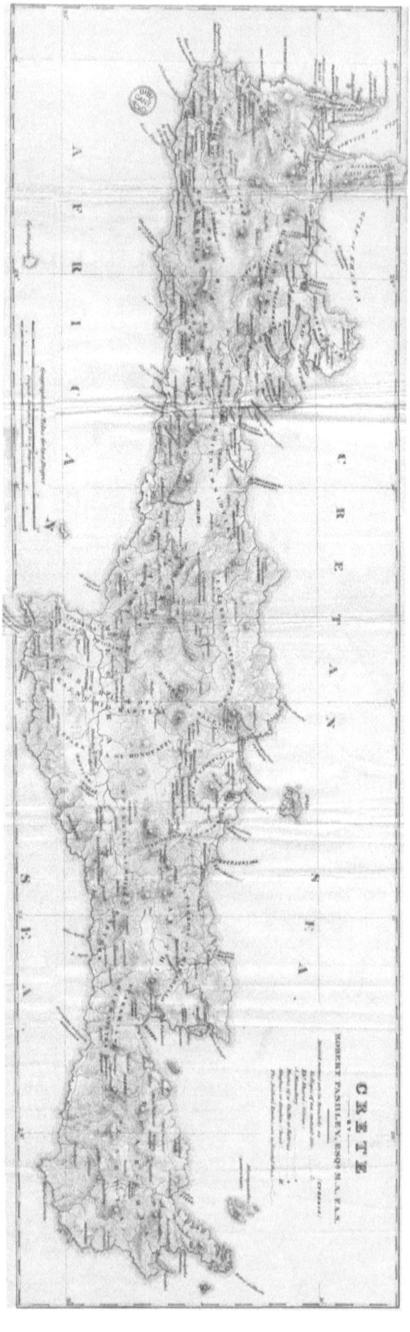

The material originally positioned here is too large for reproduction in this reissue. A PDF can be downloaded from the web address given on page iv of this book, by clicking on 'Resources Available'.

For EU product safety concerns, contact us at Calle de José Abascal, 56–1°,
28003 Madrid, Spain or eugpsr@cambridge.org.

www.ingramcontent.com/pod-product-compliance
Ingram Content Group UK Ltd.
Pitfield, Milton Keynes, MK11 3LW, UK
UKHW031623290626
13445UKWH00026B/225
* 9 7 8 1 1 0 8 0 5 0 8 3 8 *